© 2024 by BINISH SHAH. All rights reserved.

This book, titled **"Harmony of Knowledge: Science in Islamic Civilization"**, along with its contents encompassing text, illustrations, images, diagrams, and other creative elements, is the exclusive property of BINISH SHAH and is safeguarded by copyright law.

BINISH SHAH asserts full ownership and retains all rights to this book. No part of this publication may be reproduced, distributed, or transmitted in any form or by any means, such as photocopying, recording, or electronic methods, without prior written consent from the copyright holder. Brief quotations in critical reviews and certain noncommercial uses permitted by copyright law are exceptions.

This copyright notice applies to all editions, formats, and translations of the book, whether in print, digital, or any other medium or technology existing now or developed in the future. Unauthorized use or infringement may result in legal action and pursuit of remedies under applicable copyright laws.

While efforts have been made to ensure accuracy and reliability, BINISH SHAH does not guarantee the completeness or suitability of the information. Readers are responsible for evaluating and using the content judiciously.

BINISH SHAH reserves the right to make changes, updates, or corrections to the book without prior notice. Inclusion of third-party materials or references does not imply endorsement or affiliation unless used under fair use principles or with proper permissions and attributions.

For permissions, inquiries, or requests regarding the book's use, please contact BINISH SHAH through official channels listed on their Amazon author page or provided email address.

This comprehensive copyright notice serves to protect BINISH SHAH's intellectual property rights, maintain content control, and inform users about associated restrictions and permissions.

Warm regards,

BINISH SHAH

HARMONEY OF KNOWLEDGE
SCIENCE IN ISLAMIC CIVILIZATION

Table of Content

Preface

Chapter 1: The Golden Age of Islam

Chapter 2: The Quran and Scientific Inquiry

Chapter 3: Early Muslim Scholars and Their Contributions

Chapter 4: Alchemy and Chemistry in Islamic Science

Chapter 5: Astronomy and the Muslim Astronomers

Chapter 6: Mathematics and the Development of Algebra

Chapter 7: Medicine and the Muslim Physicians

Chapter 8: Geography and Cartography in Islamic Civilization

Chapter 9: Engineering and Architecture in Islamic Heritage

Chapter 10: Agriculture and Agronomy in Islamic Civilization

Chapter 11: Music and Acoustics in Islamic Culture

Chapter 12: Optics and Light in Islamic Science

Chapter 13: The Scientific Method in Islamic Scholarship

Chapter 14: Women in Science in Islamic Civilization

Chapter 15: Scientific Instruments and their Evolution

Chapter 16: Trade and Commerce in the Islamic World

Chapter 17: The Transmission of Greek Knowledge to the West

Chapter 18: Islamic Contributions to Botany and Ecology

Chapter 19: Islamic Influence on European Renaissance

Chapter 20: Islamic Science in Andalusia

Chapter 21: Islamic Ethics and Science

Chapter 22: The Role of Universities in Islamic Science

Chapter 23: Islamic Cosmology and the Universe

Chapter 24: Arabic Language and Scientific Terminology

Chapter 25: Islamic Art and Its Influence on Science

Chapter 26: Astronomy and Timekeeping in Islamic Civilization

Chapter 27: Islamic Contributions to Psychology and Psychiatry

Chapter 28: The Legacy of Al-Andalus in Science

Chapter 29: Islamic Influence on Modern Medicine

Chapter 30: The Decline of Islamic Science

Chapter 31: Revival of Interest in Islamic Science

Chapter 32: Modern Interpretations of Islamic Science

Chapter 33: Islamic Ethics and Biotechnology

Chapter 34: Islamic Science and Environmental Conservation

Chapter 35: Islamic Mathematics and Modern Cryptography

Chapter 36: Islamic Medicine and Modern Healthcare

Chapter 37: Islamic Architecture and Modern Engineering

Chapter 38: Islamic Astronomy and Modern Space Exploration

Chapter 39: Islamic Science and Globalization

Chapter 40: The Future of Islamic Science

Chapter 41: Challenges and Opportunities in Islamic Science

Chapter 42: The Role of Islamic Science in Interfaith Dialogue

Chapter 43: Islamic Science and Ethics in Artificial Intelligence

Chapter 44: Islamic Science and Sustainable Development Goals

Chapter 45: Islamic Science and Climate Change

Chapter 46: Islamic Science and the Digital Age

Chapter 47: Islamic Science and Social Justice

Chapter 48: Islamic Science and Education

Chapter 49: Islamic Science and the Quest for Knowledge

Chapter 50: Conclusion: A Vision for the Future of Science in Islamic Civilization

About the author

Preface

"Harmony of Knowledge: Science in Islamic Civilization" explores the remarkable contributions of Islamic science to the development of human knowledge and civilization. This book is a tribute to the rich heritage of Islamic science, which has played a pivotal role in shaping our understanding of the natural world and the universe.

The concept of harmony is central to Islamic science, reflecting the interconnectedness of knowledge and the unity of the universe. Islamic scholars and scientists have long recognized the harmony between the physical and spiritual realms, seeking to understand the natural world as a manifestation of divine order and wisdom.

In this book, we delve into the history, philosophy, and practices of Islamic science, exploring its key achievements and its enduring legacy. We examine the Golden Age of Islam, a period of remarkable intellectual and scientific flourishing, and we highlight the advancements made by early Muslim scholars and scientists in fields such as astronomy, mathematics, medicine, and philosophy.

We also explore the role of the Quran and Islamic ethics in shaping scientific inquiry, highlighting the importance of ethical conduct and responsible stewardship of the Earth. We discuss the contributions of Muslim scholars to various scientific disciplines, including alchemy, chemistry, astronomy, mathematics, medicine, and geography, showcasing the breadth and depth of Islamic scientific achievements.

Furthermore, we examine the transmission of Greek knowledge to the West through Islamic civilization, highlighting the role of Islamic scholars in preserving and translating ancient texts. We discuss the influence of Islamic science on the European Renaissance and the development of modern science, demonstrating the enduring impact of Islamic thought on the course of human history.

Finally, we envision the future of Islamic science, discussing its contemporary relevance and its potential to address global challenges such as climate change, social justice, and sustainable development. We believe that Islamic science has much to offer the world today, inspiring future generations of scientists and scholars to continue the quest for knowledge and understanding.

We hope that this book will serve as a valuable resource for scholars, students, and anyone interested in the fascinating story of Islamic science. May it inspire curiosity, spark dialogue, and deepen our appreciation for the beauty and harmony of the universe.

Chapter 1: The Golden Age of Islam

During the 8th to 14th centuries, the Islamic world experienced a period of remarkable intellectual and cultural flourishing known as the Golden Age of Islam. This era was characterized by significant advancements in various fields, including science, mathematics, medicine, philosophy, art, and architecture. The Golden Age was marked by a spirit of curiosity, inquiry, and open-mindedness, which allowed for the synthesis and expansion of knowledge from different cultures and civilizations.

One of the key factors that contributed to the Golden Age was the translation movement, initiated by Caliph al-Ma'mun in the 9th century. This movement aimed to translate works from Greek, Persian, and Indian sources into Arabic, making them accessible to scholars in the Islamic world. These translations played a crucial role in the transmission of knowledge and ideas, particularly in the fields of philosophy, science, and medicine.

The Golden Age saw significant advancements in the field of astronomy. Muslim astronomers made groundbreaking discoveries, such as the development of accurate astronomical instruments, the calculation of the Earth's circumference, and the formulation of models to explain the motion of celestial bodies. Scholars like Al-Battani, Al-Biruni, and Al-Farghani made significant contributions to the field, which laid the foundation for later developments in astronomy.

In the field of mathematics, Islamic scholars built upon the work of ancient Greek and Indian mathematicians to develop new concepts and techniques. The most notable achievement was the development of algebra, which was named after the Arabic word "al-jabr" meaning "reunion of broken parts." Mathematicians like Al-Khwarizmi, Al-Kindi, and Omar Khayyam made significant contributions to the field, which revolutionized the way mathematics was studied and understood.

Islamic civilization also made significant advancements in the field of medicine. Muslim physicians like Al-Razi (Rhazes) and Ibn Sina (Avicenna) made significant contributions to the field, including the development of medical encyclopedias and the establishment of hospitals with specialized wards for different diseases. These advancements not only improved medical practices in the Islamic world but also had a lasting impact on the development of medicine in Europe.

The Golden Age of Islam was not only a period of scientific and intellectual advancement but also a time of cultural and artistic flourishing. Islamic art and architecture flourished during this period, with the construction of magnificent mosques, palaces, and public buildings that showcased intricate geometric patterns, arabesques, and calligraphy.

In conclusion, the Golden Age of Islam was a period of remarkable achievement and innovation that laid the foundation for many of the scientific, mathematical, and cultural developments that shaped the modern world. It was a time when scholars from different cultures and backgrounds came together to exchange ideas, challenge existing beliefs, and push the boundaries of knowledge, leaving a lasting legacy that continues to inspire us today.

Chapter 2: The Quran and Scientific Inquiry

The Quran, the holy book of Islam, is considered by Muslims to be the literal word of God as revealed to the Prophet Muhammad. It is revered not only for its spiritual guidance but also for its perceived compatibility with scientific knowledge. Throughout history, many Muslims and scholars have interpreted certain verses of the Quran as containing scientific insights and truths that were only later discovered by modern science. This has led to a field of study known as "Islamic science" or "Quranic science," which seeks to find scientific evidence in the Quran.

One of the key aspects of the Quran that is often cited in relation to science is its emphasis on observation and reflection. The Quran repeatedly urges believers to contemplate the natural world as a means of understanding God's creation. Verses such as "Do they not look at the camels, how they are created?" (Quran 88:17) and "We will show them Our signs in the horizons and within themselves until it becomes clear to them that it is the truth" (Quran 41:53) are often cited as examples of this emphasis on observation and reflection.

Another aspect of the Quran that is often highlighted in relation to science is its description of natural phenomena. The Quran contains numerous references to natural phenomena, such as the creation of the universe, the water cycle, the development of the embryo, and the orbits of the planets, which some interpret as reflecting scientific knowledge that was not known at the time of the Quran's revelation.

For example, the Quran describes the creation of the universe in a way that is consistent with the Big Bang theory, stating, "Do not those who disbelieve see that the heavens and the earth were a closed-up mass, then We opened them out?" (Quran 21:30). Similarly, the Quran describes the development of the embryo in the womb in a way that is consistent with modern embryology, stating, "Then We made the drop into a clot, then We made the clot into a lump, then We made the lump into bones, then We clothed the bones with flesh" (Quran 23:14).

However, it is important to note that while these interpretations are compelling to some, they are not universally accepted among Muslims or scholars. The relationship between the Quran and science is complex and subject to interpretation, and there is ongoing debate among Muslims and scholars about the extent to which the Quran can be seen as containing scientific truths.

In conclusion, the Quran's emphasis on observation, reflection, and the natural world has led many Muslims and scholars to interpret certain verses as containing scientific insights. While these interpretations are not universally accepted, they reflect a belief among some that the Quran and science are compatible and that the Quran contains knowledge that was only later discovered by modern science.

Chapter 3: Early Muslim Scholars and Their Contributions

The early Muslim scholars made significant contributions to various fields of knowledge, laying the foundation for the scientific and intellectual advancements of the Islamic Golden Age. These scholars hailed from diverse backgrounds and regions, including the Arabian Peninsula, Persia, and Central Asia, and their work spanned a wide range of disciplines, including astronomy, mathematics, medicine, philosophy, and literature.

One of the most renowned early Muslim scholars was Al-Kindi (c. 801–873 CE), often referred to as the "Philosopher of the Arabs." Al-Kindi made significant contributions to philosophy, mathematics, and medicine. He was one of the first scholars to translate Greek philosophical works into Arabic, helping to introduce Greek thought to the Islamic world. Al-Kindi also made important contributions to the field of cryptography, developing methods for encrypting and decrypting messages.

Another influential scholar was Al-Farabi (c. 872–950 CE), known for his work in philosophy and political science. Al-Farabi's writings combined elements of Greek philosophy with Islamic thought, and he made significant contributions to fields such as logic, metaphysics, ethics, and music theory. His work had a profound influence on later Islamic philosophers and scholars.

One of the most celebrated early Muslim scholars was Ibn Sina (Avicenna) (980–1037 CE), whose contributions to medicine and philosophy were groundbreaking. Ibn Sina's medical encyclopedia, the "Canon of Medicine," became a standard medical textbook in Europe and the Islamic world for centuries. He also made important contributions to philosophy, logic, mathematics, and astronomy.

In the field of astronomy, early Muslim scholars like Al-Battani (c. 858–929 CE) and Al-Farghani (c. 830–870 CE) made significant advancements. Al-Battani's observations of the stars and planets led to more accurate calculations of their positions, which had practical applications in navigation and timekeeping. Al-Farghani's work on the motion of celestial bodies helped to refine existing astronomical models.

In mathematics, scholars like Al-Khwarizmi (c. 780–850 CE) and Omar Khayyam (1048–1131 CE) made important contributions. Al-Khwarizmi's work on algebra laid the foundation for the development of modern algebra, and his book "Al-Kitab al-Mukhtasar fi Hisab al-Jabr wa'l-Muqabala" (The Compendious Book on Calculation by Completion and Balancing) introduced the concept of algebra to the Islamic world and Europe. Omar Khayyam made significant contributions to algebra, geometry, and the development of the Persian calendar.

In conclusion, the early Muslim scholars made significant contributions to a wide range of fields, including philosophy, mathematics, medicine, and astronomy. Their work laid the foundation for the scientific and intellectual advancements of the Islamic Golden Age and had a lasting impact on the development of knowledge in the Islamic world and beyond.

Chapter 4: Alchemy and Chemistry in Islamic Science

Alchemy and chemistry played significant roles in the development of science in Islamic civilization. While alchemy was primarily concerned with the transformation of matter, including attempts to turn base metals into gold and discover the elixir of life, chemistry focused more on the study of substances and their properties. Islamic scholars made important contributions to both fields, laying the groundwork for the emergence of modern chemistry.

The roots of Islamic alchemy can be traced back to ancient Greek, Egyptian, and Indian traditions, which were further developed and synthesized by Muslim scholars. Alchemy was not only a quest for material wealth but also a spiritual and philosophical pursuit, with alchemists seeking to understand the nature of matter and the secrets of the universe.

One of the most influential Islamic alchemists was Jabir ibn Hayyan (Geber) (c. 721–815 CE), often considered the father of Islamic alchemy. Jabir made significant advancements in the field, including the development of laboratory equipment and techniques for chemical experimentation. He also introduced the concept of the "sulfur-mercury theory," which posited that all metals were composed of varying proportions of sulfur and mercury.

Islamic alchemists also made important contributions to the field of medicine. They developed new drugs and medicines, as well as techniques for distillation and extraction that are still used in pharmacy today. Al-Razi (Rhazes) (865–925 CE), for example, wrote extensively on alchemy and chemistry, and his works had a significant impact on the development of pharmaceutical practices in the Islamic world and Europe.

In addition to alchemy, Islamic scholars made important contributions to the field of chemistry. They conducted experiments, recorded their findings, and developed theories about the nature of matter. One of the most important works in Islamic chemistry is the "Book of Chemical Secrets" by Muhammad ibn Zakariya al-Razi (Rhazes), which contains detailed descriptions of chemical experiments and laboratory techniques.

Islamic chemists also made significant advancements in the field of metallurgy. They developed new techniques for extracting metals from ores and for producing alloys with specific properties. These advancements had practical applications in various industries, including jewelry making, metalworking, and weapon manufacturing.

In conclusion, alchemy and chemistry played significant roles in the development of science in Islamic civilization. Islamic scholars made important contributions to both fields, laying the groundwork for the emergence of modern chemistry. Their work not only advanced our understanding of the natural world but also had practical applications that benefited society as a whole.

Chapter 5: Astronomy and the Muslim Astronomers

Astronomy was a central field of study in Islamic civilization, driven by both religious and practical considerations. Muslim astronomers made significant contributions to the field, advancing knowledge in areas such as celestial mapping, observational astronomy, and astronomical theory. Their work laid the foundation for many of the astronomical advancements that would later influence European astronomy during the Renaissance.

One of the key figures in Islamic astronomy was Al-Battani (c. 858–929 CE), also known as Albategnius in the West. Al-Battani made accurate observations of the positions of stars and planets, which allowed him to refine existing astronomical models. His work, particularly his observations of the solar year, had a significant impact on later European astronomers, including Copernicus.

Another influential astronomer was Al-Farghani (c. 830–870 CE), also known as Alfraganus. Al-Farghani wrote a popular astronomical handbook called the "Elements of Astronomy," which was widely used in both the Islamic world and Europe for centuries. The book covered topics such as the motion of celestial bodies, the determination of planetary positions, and the calculation of eclipses.

Islamic astronomers also made important advancements in the field of trigonometry, which was essential for making accurate astronomical calculations. Al-Battani and Muhammad ibn Jābir al-Harani al-Battani (c. 850–929 CE), known as Alhazen in the West, both made significant contributions to trigonometry, developing new methods for calculating the positions of celestial bodies.

One of the most significant contributions of Islamic astronomy was the development of astronomical instruments. Muslim astronomers invented many of the instruments used in observational astronomy, including the astrolabe, the quadrant, and the armillary sphere. These instruments allowed astronomers to make more accurate measurements of celestial phenomena and helped to advance the field of astronomy.

Islamic astronomers also made important contributions to theoretical astronomy. They developed new models to explain the motion of celestial bodies, including the sun, moon, and planets. One of the most influential models was developed by Al-Battani, who proposed a new theory of lunar and solar motion that was more accurate than previous models.

In conclusion, Islamic astronomers made significant contributions to the field of astronomy, advancing knowledge in areas such as observational astronomy, theoretical astronomy, and astronomical instrumentation. Their work laid the foundation for many of the astronomical advancements that would later influence European astronomy, making a lasting impact on the field of astronomy as a whole.

Chapter 6: Mathematics and the Development of Algebra

Mathematics was a thriving field in Islamic civilization, with scholars making significant advancements in various branches of mathematics, including arithmetic, geometry, trigonometry, and algebra. One of the most notable achievements of Islamic mathematicians was the development of algebra, which had a profound impact on the field of mathematics and laid the foundation for modern algebraic notation and methods.

The word "algebra" itself is derived from the Arabic word "al-jabr," which means "reunion of broken parts." The term was popularized by the mathematician Al-Khwarizmi (c. 780–850 CE) in his seminal work "Kitab al-Jabr wa-l-Muqabala" (The Compendious Book on Calculation by Completion and Balancing). Al-Khwarizmi's book, written around 820 CE, was one of the first systematic treatments of algebra and introduced the basic concepts and methods that are still used in algebra today.

One of the key innovations of Islamic algebra was the use of symbols and letters to represent unknown quantities, a practice that greatly simplified the solving of algebraic equations. Al-Khwarizmi and other Islamic mathematicians developed a symbolic notation for algebraic expressions and equations, which allowed them to solve complex problems that were previously difficult or impossible to solve.

Another important development in Islamic algebra was the solution of quadratic equations. Islamic mathematicians developed methods for solving quadratic equations, including the use of geometric proofs and graphical representations. Al-Khwarizmi, Al-Khujandi (c. 940–1000 CE), and Omar Khayyam (1048–1131 CE) were among the mathematicians who made significant contributions to the solution of quadratic equations.

Islamic mathematicians also made advancements in other areas of mathematics, such as geometry and trigonometry. They made important contributions to the development of trigonometry, including the calculation of trigonometric functions and the development of trigonometric tables. Muslim mathematicians also made significant advancements in geometry, including the development of new theorems and methods for calculating areas and volumes.

In conclusion, Islamic mathematicians made significant contributions to the field of mathematics, particularly in the development of algebra. Their work laid the foundation for many of the mathematical advancements that would later influence European mathematics during the Renaissance. The development of algebra in Islamic civilization was a watershed moment in the history of mathematics, marking the beginning of a new era of mathematical inquiry and discovery.

Chapter 7: Medicine and the Muslim Physicians

Medicine was a highly developed field in Islamic civilization, building on the knowledge of ancient Greek, Roman, Persian, and Indian traditions. Muslim physicians made significant advancements in medical theory, practice, and education, laying the foundation for modern medicine.

One of the most influential figures in Islamic medicine was Al-Razi (Rhazes) (865–925 CE), who is considered one of the greatest physicians in history. Al-Razi made significant contributions to various fields of medicine, including pediatrics, obstetrics, and ophthalmology. He was also a prolific writer, producing over 200 books and treatises on medicine and related topics. His most famous work, the "Al-Hawi fi al-Tibb" (Comprehensive Book on Medicine), was a comprehensive medical encyclopedia that became a standard reference in both the Islamic world and Europe.

Another prominent physician was Ibn Sina (Avicenna) (980–1037 CE), whose medical encyclopedia, the "Canon of Medicine," became a standard medical textbook in Europe and the Islamic world for centuries. Ibn Sina made significant contributions to medical theory, including the development of a comprehensive system of medicine based on the principles of Greek philosophy and Islamic theology. He also made important advancements in pharmacology and the classification of diseases.

Muslim physicians also made important contributions to the field of surgery. Al-Zahrawi (Albucasis) (936–1013 CE) was a pioneering surgeon who wrote the "Kitab al-Tasrif" (Book of Medical Knowledge), which became the most widely used surgical textbook in Europe for over 500 years. Al-Zahrawi's work included detailed descriptions of surgical techniques, instruments, and treatments for various conditions.

In addition to their medical advancements, Muslim physicians also made important contributions to public health. They developed new methods for sanitation and hygiene, including the use of clean water, sewage systems, and public baths. They also established hospitals with specialized wards for different diseases, which provided medical care to patients regardless of their social or economic status.

Islamic medicine was not limited to the work of individual physicians; it was also supported by a system of medical education and research. Islamic scholars established medical schools and hospitals where students could receive comprehensive training in medicine, surgery, and pharmacology. These institutions played a crucial role in the dissemination of medical knowledge and the advancement of medical science.

In conclusion, medicine was a highly developed field in Islamic civilization, thanks to the contributions of Muslim physicians. Their advancements in medical theory, practice, and education laid the foundation for modern medicine and had a lasting impact on the development of healthcare around the world.

Chapter 8: Geography and Cartography in Islamic Civilization

Geography and cartography were important fields of study in Islamic civilization, driven by both practical and intellectual interests. Muslim geographers and cartographers made significant advancements in the understanding of the world's geography, the creation of accurate maps, and the development of navigational techniques.

One of the most influential figures in Islamic geography was Al-Masudi (c. 896–956 CE), often referred to as the "Herodotus of the Arabs." Al-Masudi wrote several works on geography and history, including the famous "Muruj adh-Dhahab wa-Ma'adin al-Jawhar" (The Meadows of Gold and Mines of Gems), which is considered one of the greatest works of medieval geography. In his writings, Al-Masudi described the geography, history, and cultures of the known world, from Africa and Europe to Asia and the Far East.

Another important geographer was Ibn Khaldun (1332–1406 CE), whose work "Muqaddimah" (Introduction to History) included a detailed discussion of geography and the environment's influence on human societies. Ibn Khaldun's work laid the foundation for modern historiography and sociology and had a significant impact on European thought during the Renaissance.

Islamic cartographers also made significant advancements in the field of mapmaking. One of the most famous Islamic maps is the "Tabula Rogeriana" (The Book of Roger), created by the Andalusian cartographer Al-Idrisi in 1154 CE for the Norman king Roger II of Sicily. The map was remarkable for its accuracy and level of detail, depicting the known world at the time with remarkable precision.

Muslim cartographers also made important advancements in the development of navigational techniques. They developed new methods for determining latitude and longitude, including the use of astrolabes and celestial navigation. These advancements were crucial for the development of global trade and exploration.

Islamic geography and cartography were not only practical endeavors but also intellectual pursuits. Muslim scholars were interested in understanding the world's geography as a reflection of God's creation and the diversity of human societies. Their work laid the foundation for modern geography and cartography, influencing European scholars during the Renaissance and beyond.

In conclusion, geography and cartography were important fields of study in Islamic civilization, with Muslim scholars making significant advancements in the understanding of the world's geography and the creation of accurate maps. Their work had a lasting impact on the development of geography and cartography and played a crucial role in shaping our understanding of the world.

Chapter 9: Engineering and Architecture in Islamic Heritage

Engineering and architecture were highly developed fields in Islamic civilization, characterized by innovative designs, advanced construction techniques, and the creation of iconic structures that still stand today. Muslim engineers and architects made significant advancements in various aspects of building design, construction, and urban planning, leaving a lasting legacy that continues to influence architecture around the world.

One of the most notable achievements of Islamic engineering was the development of advanced water management systems. Muslim engineers built elaborate networks of aqueducts, canals, and reservoirs to transport and store water for irrigation, drinking, and sanitation. These systems were crucial for the development of agriculture, cities, and civilization in arid regions.

One of the most famous examples of Islamic water engineering is the Alhambra in Granada, Spain. The Alhambra features a complex system of water channels, fountains, and pools that not only provide water for the palace but also create a tranquil and beautiful environment that is an integral part of the architectural design.

Islamic architecture is also known for its distinctive style, characterized by the extensive use of geometric patterns, arabesques, and calligraphy. Islamic architects used these elements to create intricate designs that adorned mosques, palaces, and public buildings, reflecting the Islamic belief in the unity and harmony of all creation.

One of the most iconic examples of Islamic architecture is the Great Mosque of Cordoba in Spain. The mosque is renowned for its horseshoe arches, intricate geometric patterns, and decorative tilework, which combine to create a stunning visual effect that is both grand and harmonious.

Islamic engineers and architects also made important advancements in structural engineering. They developed new techniques for building domes, arches, and vaults, which allowed for the creation of larger and more complex structures. One of the most famous examples of Islamic structural engineering is the dome of the Hagia Sophia in Istanbul, Turkey, which was built in the 6th century and remains one of the largest domes in the world.

In addition to their architectural achievements, Islamic engineers also made significant advancements in other fields of engineering, including mechanical engineering, civil engineering, and military engineering. Muslim engineers developed new technologies for irrigation, transportation, and warfare, which had a profound impact on the development of technology and civilization.

In conclusion, engineering and architecture were highly developed fields in Islamic civilization, characterized by innovative designs, advanced construction techniques, and the creation of iconic structures. Muslim engineers and architects made significant advancements in various aspects of building design, construction, and urban planning, leaving a lasting legacy that continues to influence architecture around the world.

Chapter 10: Agriculture and Agronomy in Islamic Civilization

Agriculture was a vital component of Islamic civilization, providing sustenance for urban and rural populations and supporting economic development and cultural exchange. Muslim farmers and agronomists made significant advancements in agricultural practices, irrigation techniques, crop cultivation, and land management, which had a lasting impact on agriculture in the Islamic world and beyond.

One of the key achievements of Islamic agriculture was the development of sophisticated irrigation systems. Muslim engineers built elaborate networks of canals, aqueducts, and reservoirs to bring water to arid and semi-arid regions, enabling agriculture to thrive in areas that were previously uninhabitable. These irrigation systems were crucial for the development of agriculture and the growth of cities in regions such as Spain, North Africa, the Middle East, and Central Asia.

Islamic agronomists also made important contributions to crop cultivation and land management. They developed new techniques for crop rotation, soil conservation, and pest control, which helped to improve agricultural productivity and sustainability. Muslim agronomists also conducted experiments to determine the best methods for growing crops in different environments, leading to advancements in plant breeding and agricultural science.

One of the most famous Islamic agronomists was Ibn Al-Awwam (12th century CE), whose work "Kitab al-Filaha"

(Book on Agriculture) became a standard reference in the Islamic world and Europe for centuries. Ibn Al-Awwam's book covered a wide range of topics related to agriculture, including crop cultivation, irrigation, soil management, and pest control. It also included detailed instructions on how to grow a variety of crops, fruits, and vegetables, making it an invaluable resource for farmers.

Islamic agriculture was also characterized by the cultivation of new crops and the introduction of agricultural innovations from other regions. Muslim traders and travelers brought new crops and agricultural techniques from Africa, Asia, and Europe, which were adapted and integrated into Islamic agricultural practices. This exchange of agricultural knowledge and practices helped to improve agricultural productivity and diversity in the Islamic world.

In conclusion, agriculture was a vital component of Islamic civilization, supported by advancements in irrigation, crop cultivation, and land management. Muslim farmers and agronomists made significant contributions to agricultural science and technology, which had a lasting impact on agriculture in the Islamic world and beyond. Their innovations helped to improve agricultural productivity, sustainability, and diversity, ensuring the continued growth and prosperity of Islamic civilization.

Chapter 11: Music and Acoustics in Islamic Culture

Music has been an integral part of Islamic culture, with a rich tradition that spans centuries and encompasses a wide variety of musical styles, instruments, and genres. Islamic music has been influenced by diverse cultural and regional traditions, resulting in a vibrant and diverse musical heritage.

One of the key features of Islamic music is its use of maqamat, which are melodic modes or scales. Maqamat are used to create melodies and improvisations in Islamic music, and they vary in complexity and mood. Each maqam has its own set of rules and conventions, which govern how it is performed and how it evokes different emotions and feelings.

Islamic music also places a strong emphasis on rhythm and percussion instruments. Percussion instruments such as the tabla, daf, and darbuka are commonly used in Islamic music to create rhythmic patterns and accompaniments. These instruments are often played in conjunction with melodic instruments such as the oud, qanun, and ney, which create the melody and harmony of the music.

Acoustics played a crucial role in the development of Islamic music, particularly in the design and construction of mosques and other religious buildings. Muslim architects and engineers developed innovative techniques for shaping acoustics in mosques to enhance the sound of the call to prayer (adhan) and create a reverberant and harmonious acoustic environment for worshipers.

One of the most famous examples of Islamic acoustics is the Great Mosque of Cordoba in Spain. The mosque's design features a series of horseshoe arches and columns that create a unique acoustic environment, amplifying the sound of the imam's voice during prayers and creating a sense of unity and harmony among worshipers.

Islamic music also played a role in the development of Western music. During the Middle Ages, Islamic Spain was a center of cultural exchange between the Islamic world and Europe, and many musical instruments and musical concepts were introduced to Europe through contact with Islamic culture. The lute, for example, was introduced to Europe from the Islamic world and became a popular instrument in Western music.

In conclusion, music and acoustics have played a significant role in Islamic culture, with a rich tradition that has influenced music around the world. Islamic music is characterized by its use of maqamat, emphasis on rhythm and percussion, and innovative approaches to acoustics. Islamic music continues to be a vibrant and evolving art form, reflecting the diversity and richness of Islamic culture.

Chapter 12: Optics and Light in Islamic Science

Optics, the study of light and its behavior, was a field of great interest and advancement in Islamic science. Muslim scholars made significant contributions to the understanding of optics, light, and vision, which had a profound impact on fields such as astronomy, physics, and philosophy.

One of the key figures in Islamic optics was Al-Kindi (c. 801–873 CE), often referred to as the "Philosopher of the Arabs." Al-Kindi made important contributions to optics, including the development of theories on vision, light, and the nature of light rays. He also wrote extensively on the properties of light and its behavior, laying the foundation for later developments in optics.

Another influential figure was Ibn al-Haytham (Alhazen) (965–1040 CE), whose work on optics, particularly his book "Kitab al-Manazir" (Book of Optics), had a profound impact on the development of the field. Ibn al-Haytham made significant advancements in the study of vision, including the discovery that light enters the eye and is not emitted from it, as was previously believed. He also developed a theory of visual perception based on the concept of rays of light reflecting off objects and entering the eye.

Islamic scholars also made important advancements in the field of lenses and optics. Ibn Sahl (c. 940–1000 CE) developed the first mathematical theory of lenses, which laid the foundation for the development of modern optics. Muslim engineers and architects also used lenses and mirrors in innovative ways, such as in the design of astronomical instruments and the creation of intricate patterns of light and shadow in architecture.

Islamic optics also had a significant impact on European thought and science during the Middle Ages. The works of Muslim scholars such as Al-Kindi, Ibn al-Haytham, and Ibn Sahl were translated into Latin and influenced European scholars such as Roger Bacon and Leonardo da Vinci. The translation of Islamic works on optics played a crucial role in the development of Western science and the Renaissance.

In conclusion, optics and light were important fields of study in Islamic science, with Muslim scholars making significant advancements in the understanding of light, vision, and optics. Their work laid the foundation for later developments in optics and had a lasting impact on the fields of astronomy, physics, and philosophy.

Chapter 13: The Scientific Method in Islamic Scholarship

The scientific method, which involves systematic observation, measurement, experimentation, and the formulation, testing, and modification of hypotheses, was a fundamental aspect of Islamic scholarship. Muslim scientists and scholars developed and refined the scientific method, making significant contributions to the advancement of knowledge in fields such as astronomy, mathematics, medicine, and philosophy.

One of the key principles of the scientific method is empirical observation, which involves gathering data through direct observation and measurement. Muslim scientists were known for their meticulous observations of the natural world, which laid the foundation for many scientific discoveries and advancements. For example, Muslim astronomers made precise observations of the positions of stars and planets, which allowed them to develop accurate models of the heavens.

Another important aspect of the scientific method is experimentation, which involves testing hypotheses through controlled experiments. Muslim scientists were pioneers in the field of experimental science, conducting experiments to test their theories and hypotheses. Al-Razi, for example, conducted experiments to test the properties of various substances and the effects of different treatments on patients.

Muslim scholars also emphasized the importance of rational inquiry and critical thinking in the pursuit of knowledge. They were known for their rigorous approach to scholarship, which involved questioning established beliefs and theories and seeking evidence to support their conclusions. This approach to scholarship was instrumental in the development of the scientific method and the advancement of knowledge in Islamic civilization.

One of the key features of the scientific method is its emphasis on collaboration and communication among scientists. Muslim scholars were known for their exchange of ideas and knowledge across cultural and geographic boundaries, which helped to advance scientific knowledge and understanding. Muslim scholars translated and preserved ancient Greek, Roman, Persian, and Indian scientific texts, making them available to scholars in the Islamic world and Europe.

In conclusion, the scientific method was a fundamental aspect of Islamic scholarship, with Muslim scientists and scholars making significant contributions to its development and refinement. Their emphasis on empirical observation, experimentation, rational inquiry, and collaboration laid the foundation for many scientific advancements and helped to shape the course of scientific inquiry for centuries to come.

Chapter 14: Women in Science in Islamic Civilization

Women played a significant role in the advancement of science in Islamic civilization, contributing to various fields such as medicine, astronomy, mathematics, and philosophy. Despite the challenges they faced in a male-dominated society, many women pursued careers in science and made important contributions to their respective fields.

One notable example is Mariam al-Ijliya al-Astrulabi, a 10th-century astronomer and maker of astrolabes. She was known for her expertise in astronomy and astrolabe-making, a complex instrument used for astronomical measurements and timekeeping. Her astrolabes were renowned for their accuracy and craftsmanship, and she was highly respected for her contributions to the field of astronomy.

Another influential woman in Islamic science was Fatima al-Fihri, who founded the University of al-Qarawiyyin in Fez, Morocco, in 859 CE. The university became one of the leading centers of learning in the Islamic world and played a crucial role in the preservation and transmission of knowledge, including scientific knowledge, during the Middle Ages.

Women also made important contributions to the field of medicine in Islamic civilization. Rufaida al-Aslamia, for example, was a renowned nurse and herbalist who lived in the 7th century and is considered one of the first Muslim nurses. She is credited with establishing the first documented hospital in Islam and was known for her compassion and skill in treating patients.

In the field of mathematics, Islamic women made significant advancements in areas such as algebra and geometry. Sutayta al-Mahamili, a 10th-century mathematician, wrote a commentary on the mathematical works of Al-Khwarizmi, one of the leading mathematicians of the Islamic Golden Age. Her commentary helped to popularize Al-Khwarizmi's work and contributed to the spread of algebra in the Islamic world and Europe.

Despite their contributions, women in Islamic science often faced societal and cultural barriers that limited their opportunities for education and professional advancement. However, many women were able to overcome these challenges and make lasting contributions to science, thanks to their perseverance, talent, and dedication to their fields.

In conclusion, women played a significant role in the advancement of science in Islamic civilization, making important contributions to fields such as astronomy, medicine, mathematics, and philosophy. Their achievements are a testament to their intellect, determination, and passion for knowledge, and their contributions have enriched our understanding of the world and its wonders.

Chapter 15: Scientific Instruments and their Evolution

Scientific instruments played a crucial role in the advancement of science in Islamic civilization, facilitating precise measurements, observations, and experiments in fields such as astronomy, optics, and medicine. Muslim scientists and engineers developed a wide variety of instruments, many of which were based on earlier Greek, Persian, Indian, and Chinese designs, but were refined and improved upon to meet the specific needs of Islamic scholars.

One of the most important scientific instruments in Islamic civilization was the astrolabe, a complex instrument used for measuring the positions of celestial objects, telling time, and solving problems related to astronomy and astrology. The astrolabe was invented in ancient Greece but was greatly refined and improved by Muslim astronomers such as Al-Battani and Ibn al-Sahl. The astrolabe became one of the most widely used instruments in astronomy and navigation for centuries.

Another important instrument was the quadrant, which was used for measuring the altitude of celestial objects. The quadrant was used in conjunction with the astrolabe and other instruments to make precise astronomical observations and calculations. Muslim astronomers such as Al-Farghani and Al-Battani made important advancements in the design and use of the quadrant, improving its accuracy and utility.

Islamic engineers and architects also developed innovative instruments for use in architecture and construction. One example is the qanat, an underground irrigation system used to transport water from aquifers to the surface for irrigation and drinking. The qanat was a marvel of engineering and helped to support agriculture and urban development in arid regions.

In the field of optics, Islamic scientists developed new instruments for studying light and vision. Ibn al-Haytham, for example, developed the camera obscura, a precursor to the modern camera, which was used to study the behavior of light and create accurate drawings of objects. The camera obscura was later used by European artists such as Leonardo da Vinci to create realistic paintings.

Islamic civilization also made important advancements in medical instruments. Muslim physicians developed new instruments for surgery, diagnosis, and treatment, including forceps, scalpels, and syringes. These instruments helped to improve the practice of medicine and the treatment of patients.

In conclusion, scientific instruments played a crucial role in the advancement of science in Islamic civilization, facilitating precise measurements, observations, and experiments in fields such as astronomy, optics, and medicine. Muslim scientists and engineers developed a wide variety of instruments, many of which were based on earlier designs but were refined and improved upon to meet the specific needs of Islamic scholars. These instruments helped to advance knowledge and understanding in a wide range of scientific disciplines and had a lasting impact on the development of science and technology.

Chapter 16: Trade and Commerce in the Islamic World

Trade and commerce played a crucial role in the Islamic world, contributing to the economic prosperity and cultural exchange that characterized Islamic civilization. Islamic societies were active participants in the global trade networks that connected Asia, Africa, and Europe, facilitating the exchange of goods, ideas, and technologies.

One of the key factors that contributed to the success of Islamic trade was the establishment of a vast network of trade routes, including the Silk Road, the Indian Ocean trade routes, and the Trans-Saharan trade routes. These trade routes connected major centers of commerce and culture, allowing for the exchange of goods such as spices, textiles, precious metals, and luxury goods.

Islamic merchants played a central role in facilitating trade along these routes, acting as intermediaries between different regions and cultures. Muslim merchants were known for their business acumen, honesty, and reliability, which helped to establish trust and facilitate trade across vast distances.

Islamic societies also developed sophisticated financial and commercial practices that supported trade and commerce. One of the most important innovations was the development of the Islamic banking system, which prohibited the charging of interest (riba) and emphasized risk-sharing and ethical investment practices. Islamic banks provided crucial financial services to merchants and traders, facilitating the flow of capital and investment across the Islamic world.

Islamic cities were also important centers of trade and commerce, with bustling marketplaces and trade guilds that regulated and facilitated trade. Cities such as Baghdad, Cairo, and Damascus became major hubs of commerce, attracting merchants and traders from across the Islamic world and beyond.

The Islamic world was also a center of manufacturing and craftsmanship, producing a wide variety of goods that were highly sought after in global markets. Islamic artisans were renowned for their skill in producing textiles, ceramics, metalwork, and other luxury goods that were prized for their quality and craftsmanship.

In addition to goods, Islamic trade also facilitated the exchange of ideas, technologies, and cultural practices. The Islamic world was a melting pot of cultures and civilizations, and trade played a crucial role in the transmission of knowledge and innovation across different regions and cultures.

In conclusion, trade and commerce were integral components of Islamic civilization, contributing to its economic prosperity, cultural exchange, and technological advancement. Islamic societies were active participants in the global trade networks of the medieval world, facilitating the exchange of goods, ideas, and technologies that enriched the fabric of Islamic culture and society.

Chapter 17: The Transmission of Greek Knowledge to the West

The transmission of Greek knowledge to the West was a pivotal event in the history of science and scholarship, shaping the intellectual development of Europe during the Middle Ages and the Renaissance. Islamic scholars played a crucial role in this transmission, preserving, translating, and expanding upon the works of ancient Greek philosophers, scientists, and mathematicians.

One of the key figures in the transmission of Greek knowledge was the Abbasid caliph Al-Ma'mun (786–833 CE), who established the House of Wisdom (Bayt al-Hikmah) in Baghdad. The House of Wisdom was a major center of learning and scholarship, where Greek, Persian, Indian, and Chinese texts were translated into Arabic. Al-Ma'mun's patronage of scholarship and translation helped to preserve many ancient Greek texts that were at risk of being lost.

One of the most important translators associated with the House of Wisdom was Hunayn ibn Ishaq (c. 809–873 CE), a Christian Arab scholar who translated many works of Greek philosophy, medicine, and science into Arabic. Hunayn's translations were renowned for their accuracy and fidelity to the original texts, making them valuable sources of knowledge for scholars in the Islamic world and later in Europe.

Another important figure in the transmission of Greek knowledge was Al-Kindi (c. 801–873 CE), often referred to as the "Philosopher of the Arabs." Al-Kindi was a prolific writer and scholar who translated many works of Greek philosophy and science into Arabic. He also wrote original works on philosophy, mathematics, and astronomy, which helped to introduce Greek ideas and methods to the Islamic world.

Islamic scholars also made important contributions to the expansion and development of Greek knowledge. Muslim philosophers such as Al-Farabi (c. 872–950 CE) and Ibn Sina (Avicenna) (980–1037 CE) synthesized Greek philosophical ideas with Islamic theology and philosophy, creating a rich and diverse intellectual tradition known as Islamic philosophy.

The transmission of Greek knowledge from the Islamic world to Europe was facilitated by the translation movement that took place during the Middle Ages. European scholars such as Gerard of Cremona (c. 1114–1187 CE) and Robert of Ketton (c. 1110–1160 CE) traveled to Islamic Spain to study Arabic texts and translate them into Latin. These translations played a crucial role in the revival of learning in Europe during the Renaissance, introducing European scholars to the works of Aristotle, Plato, Euclid, and other Greek thinkers.

In conclusion, the transmission of Greek knowledge to the West was a complex and multifaceted process that involved the efforts of Islamic scholars, translators, and European scholars. Islamic scholars played a crucial role in preserving, translating, and expanding upon the works of ancient Greek thinkers, which had a profound impact on the intellectual development of Europe and the shaping of Western civilization.

Chapter 18: Islamic Contributions to Botany and Ecology

Islamic civilization made significant contributions to the fields of botany and ecology, with Muslim scholars and scientists making important advancements in the study of plants, their classification, uses, and ecological significance. Islamic scholars were among the first to systematically study plants and their properties, laying the foundation for modern botany and ecology.

One of the key figures in Islamic botany was Al-Dinawari (c. 815–896 CE), a Persian botanist who wrote the "Book of Plants," a comprehensive treatise on botany that classified over a thousand different plant species. Al-Dinawari's work was based on observation and experimentation, and he categorized plants based on their uses, such as food, medicine, and dyeing.

Another important contribution was made by Ibn Bassal (fl. 1085–1137 CE), a Spanish-Arab botanist who wrote the "Book of Agriculture," which included a detailed description of plant cultivation, irrigation, and soil management. Ibn Bassal's work was influential in the development of agricultural practices in Islamic Spain and beyond.

Islamic scholars also made important advancements in the study of ecology, particularly in understanding the relationships between organisms and their environment. Al-Jahiz (c. 776–869 CE), an Arab scholar, wrote extensively on the topic of ecology, discussing the interdependence of organisms and their adaptation to different environments. Al-Jahiz's work was ahead of its time and foreshadowed many modern ecological concepts.

Islamic agriculture was also characterized by the development of innovative techniques for sustainable farming and land management. Muslim farmers developed methods for soil conservation, crop rotation, and water management, which helped to improve agricultural productivity and sustainability. These techniques were based on a deep understanding of the natural world and the ecological principles that govern it.

Islamic civilization also made important contributions to the study of medicinal plants. Muslim physicians and pharmacologists compiled extensive pharmacopoeias that documented the medicinal properties of various plants and herbs. These pharmacopoeias were used as reference guides by physicians and pharmacists and helped to advance the field of herbal medicine.

In conclusion, Islamic civilization made significant contributions to the fields of botany and ecology, with Muslim scholars and scientists making important advancements in the study of plants, their classification, uses, and ecological significance. Islamic scholars were among the first to systematically study plants and their properties, laying the foundation for modern botany and ecology. Their work helped to expand our understanding of the natural world and its complex ecosystems, and their insights continue to inform and inspire scientists and researchers today.

Chapter 19: Islamic Influence on European Renaissance

The European Renaissance, a period of cultural and intellectual revival in Europe from the 14th to the 17th centuries, was significantly influenced by Islamic civilization. Islamic scholars, scientists, and philosophers played a crucial role in the transmission of knowledge and ideas from the Islamic world to Europe, sparking a renewed interest in classical learning and contributing to the development of humanism, science, and art in Europe.

One of the key areas of Islamic influence on the Renaissance was in the field of science and philosophy. Islamic scholars preserved and translated the works of ancient Greek philosophers and scientists, such as Aristotle, Plato, Euclid, and Ptolemy, whose ideas were largely lost to Europe during the Dark Ages. These translations introduced European scholars to Greek thought and laid the foundation for the development of modern science and philosophy in Europe.

Islamic scholars also made important advancements in science and technology that had a direct impact on the European Renaissance. Muslim astronomers, such as Al-Battani and Ibn al-Shatir, made significant contributions to astronomy and mathematics, which helped to advance European understanding of the cosmos. Islamic engineers and architects developed innovative techniques for building and construction, such as the use of arches, domes, and geometric patterns, which influenced European architecture during the Renaissance.

Islamic art and literature also had a profound impact on the European Renaissance. Islamic art, with its intricate geometric patterns, arabesques, and calligraphy, inspired European artists and designers, such as Leonardo da Vinci and Albrecht Dürer, who incorporated Islamic artistic elements into their own work. Islamic literature, with its rich tradition of poetry, storytelling, and philosophical writing, influenced European writers and poets, such as Dante Alighieri and Geoffrey Chaucer, who drew inspiration from Islamic literary themes and styles.

Islamic trade and commerce also played a role in shaping the European Renaissance. Muslim merchants and traders brought goods, ideas, and technologies from the Islamic world to Europe, stimulating economic growth and cultural exchange. The trade routes that connected the Islamic world with Europe facilitated the exchange of goods such as spices, textiles, and ceramics, which had a profound impact on European society and culture.

In conclusion, Islamic civilization had a significant influence on the European Renaissance, contributing to the revival of learning, science, art, and culture in Europe. Islamic scholars, scientists, and philosophers played a crucial role in transmitting knowledge and ideas from the Islamic world to Europe, sparking a renewed interest in classical learning and laying the foundation for the development of modern Europe. The Islamic influence on the European Renaissance is a testament to the richness and diversity of Islamic civilization and its enduring impact on the world.

Chapter 20: Islamic Science in Andalusia

Islamic science flourished in Andalusia (Islamic Spain) during the medieval period, contributing to significant advancements in various fields such as astronomy, mathematics, medicine, and philosophy. Andalusia, with its rich cultural and intellectual environment, became a center of learning and scholarship that attracted scholars and scientists from across the Islamic world and beyond.

One of the key features of Islamic science in Andalusia was its emphasis on the translation and preservation of ancient Greek, Roman, Persian, and Indian texts. Muslim scholars in Andalusia translated many important works into Arabic, making them accessible to scholars in the Islamic world and Europe. These translations played a crucial role in the transmission of knowledge and the revival of learning in Europe during the Renaissance.

One of the most famous centers of learning in Andalusia was the city of Cordoba, which became a major hub of intellectual and scientific activity. The city was home to the Great Mosque of Cordoba, which housed a library and a university that attracted scholars from across the Islamic world. Cordoba was also known for its astronomical observatory, where Muslim astronomers made important observations and calculations.

One of the most influential figures in Islamic science in Andalusia was Ibn Rushd (Averroes) (1126–1198 CE), a philosopher, physician, and jurist who made significant contributions to the fields of philosophy and medicine. Ibn Rushd's commentaries on the works of Aristotle were instrumental in reintroducing Aristotelian philosophy to Europe and had a profound impact on the development of Western thought.

Another important figure was Ibn Zuhr (Avenzoar) (1091–1161 CE), a physician and surgeon who made important advancements in the field of medicine. Ibn Zuhr's work on surgery and pharmacology was highly influential and helped to advance medical knowledge in Europe.

Islamic science in Andalusia also made important contributions to the field of mathematics. Muslim mathematicians in Andalusia, such as Al-Zarqali (Azarquiel) (1029–1087 CE), made significant advancements in trigonometry and astronomy, developing new instruments and techniques for measuring time and predicting celestial events.

In conclusion, Islamic science in Andalusia was characterized by its emphasis on learning, translation, and innovation. Muslim scholars and scientists in Andalusia made significant contributions to various fields of science, which had a lasting impact on the development of knowledge in Europe and the Islamic world. Andalusia's legacy of intellectual and scientific achievement continues to be celebrated today as a testament to the region's rich cultural and scientific heritage.

Chapter 21: Islamic Ethics and Science

Islamic ethics, rooted in the teachings of the Quran and the traditions of the Prophet Muhammad, played a significant role in shaping the practice of science in Islamic civilization. Islamic scholars and scientists adhered to ethical principles that emphasized the pursuit of knowledge, the importance of truthfulness, and the responsible use of scientific knowledge for the betterment of society.

One of the key ethical principles in Islamic science is the concept of "niyyah," or intention. Muslim scholars and scientists were encouraged to seek knowledge with the intention of serving humanity and seeking the pleasure of God. This ethical principle helped to guide their research and ensure that their scientific endeavors were conducted with sincerity and humility.

Another important ethical principle in Islamic science is the concept of "adl," or justice. Muslim scholars and scientists were expected to be fair and impartial in their research and to consider the ethical implications of their work. This principle helped to ensure that scientific knowledge was used for the benefit of all people and that the rights of individuals were respected.

Islamic ethics also emphasized the importance of honesty and truthfulness in scientific inquiry. Muslim scholars and scientists were expected to report their findings accurately and to avoid exaggeration or distortion of facts. This commitment to honesty helped to build trust and credibility in Islamic science and ensured that scientific knowledge was based on sound evidence and reasoning.

Islamic ethics also guided the use of scientific knowledge for practical purposes. Muslim scholars and scientists were encouraged to use their knowledge to improve the welfare of society and to alleviate human suffering. This ethical imperative led to important advancements in fields such as medicine, agriculture, and engineering, which benefited people across the Islamic world and beyond.

In conclusion, Islamic ethics played a significant role in shaping the practice of science in Islamic civilization. Muslim scholars and scientists adhered to ethical principles that emphasized the pursuit of knowledge, the importance of truthfulness, and the responsible use of scientific knowledge for the betterment of society. These ethical principles helped to ensure that Islamic science was conducted with integrity and compassion, and that its benefits were shared with all people.

Chapter 22: The Role of Universities in Islamic Science

Universities played a crucial role in the development and dissemination of scientific knowledge in Islamic civilization. Islamic universities, known as madrasas, were centers of learning and scholarship that attracted scholars and students from across the Islamic world and beyond. These universities played a key role in advancing knowledge in fields such as astronomy, mathematics, medicine, and philosophy.

One of the earliest Islamic universities was the University of al-Qarawiyyin in Fez, Morocco, founded by Fatima al-Fihri in 859 CE. The University of al-Qarawiyyin became one of the leading centers of learning in the Islamic world, with a strong focus on Islamic studies, theology, law, and philosophy. The university also played a crucial role in the preservation and transmission of knowledge, including scientific knowledge, during the Middle Ages.

Another important Islamic university was the Al-Azhar University in Cairo, Egypt, founded in 970 CE. Al-Azhar University became a major center of learning and scholarship, with a strong emphasis on Islamic studies, theology, and Arabic literature. The university also played a significant role in the development of Islamic science, particularly in the fields of astronomy, mathematics, and medicine.

Islamic universities were characterized by their commitment to academic freedom and intellectual inquiry. Scholars and students were encouraged to pursue knowledge and research in a wide range of fields, and there was a spirit of openness and tolerance towards different ideas and perspectives. This intellectual environment helped to foster creativity and innovation in Islamic science and contributed to the development of new ideas and theories.

Islamic universities also played a crucial role in the translation and dissemination of scientific knowledge. Muslim scholars in these universities translated many important works from Greek, Persian, Indian, and Chinese into Arabic, making them accessible to scholars in the Islamic world and Europe. These translations played a key role in the transmission of knowledge and the revival of learning in Europe during the Renaissance.

In conclusion, Islamic universities played a crucial role in the development and dissemination of scientific knowledge in Islamic civilization. These universities were centers of learning and scholarship that attracted scholars and students from across the Islamic world and beyond. Islamic universities played a key role in advancing knowledge in fields such as astronomy, mathematics, medicine, and philosophy, and their contributions continue to be celebrated today as a testament to the richness and diversity of Islamic scholarship.

Chapter 23: Islamic Cosmology and the Universe

Islamic cosmology is the study of the universe as understood in Islamic theology and philosophy. Islamic scholars and scientists developed a comprehensive cosmology that integrated religious beliefs with scientific observations, creating a holistic view of the universe that encompassed both the physical and metaphysical realms.

One of the key aspects of Islamic cosmology is the concept of Tawhid, or the unity of God. Islamic cosmology emphasizes that the universe is created and sustained by God and that all aspects of the universe are interconnected and dependent on each other. This concept of unity is reflected in the Islamic belief that the universe is a unified and harmonious creation of God.

Islamic cosmology also incorporates elements of Greek and Persian cosmology, particularly in the fields of astronomy and astrology. Muslim astronomers developed sophisticated models of the universe, based on a geocentric view with the Earth at the center, surrounded by concentric spheres that contained the sun, moon, planets, and stars. These models were used to predict the movements of celestial bodies and to create calendars and horoscopes.

One of the most influential figures in Islamic cosmology was Ibn al-Haytham (Alhazen) (965–1040 CE), whose work on optics and astronomy had a profound impact on the development of Islamic cosmology. Ibn al-Haytham rejected the Aristotelian view of the universe and developed a new model based on empirical observation and mathematical analysis. His work laid the foundation for later developments in astronomy and cosmology.

Islamic cosmology also incorporated elements of mysticism, particularly in the Sufi tradition. Sufi cosmology emphasizes the spiritual significance of the universe and the interconnectedness of all beings. Sufi mystics often use cosmological symbolism to convey spiritual truths and to illustrate the relationship between the individual soul and the divine.

In conclusion, Islamic cosmology is a rich and diverse field that integrates religious beliefs with scientific observations to create a holistic view of the universe. Islamic scholars and scientists developed sophisticated models of the universe that were based on both observation and revelation, creating a comprehensive cosmology that continues to inspire scholars and seekers of knowledge today.

Chapter 24: Arabic Language and Scientific Terminology

The Arabic language played a crucial role in the development and dissemination of scientific knowledge in Islamic civilization. Arabic, as the language of the Quran and the lingua franca of the Islamic world, became the medium through which scientific ideas were communicated, preserved, and transmitted across cultures and generations.

One of the key contributions of the Arabic language to science was the development of a rich and precise scientific terminology. Muslim scholars and scientists developed a comprehensive system of scientific terminology that was based on the Arabic language's unique grammatical and phonetic structure. This system of terminology was used to describe and classify scientific concepts, making it easier for scholars to communicate and share knowledge.

Arabic scientific terminology drew upon a variety of sources, including Arabic, Greek, Persian, and Sanskrit. Muslim scholars translated many important scientific works from these languages into Arabic, and in the process, they adopted and adapted many scientific terms. For example, the Arabic word "algebra" (al-jabr) is derived from the title of a book by the Persian mathematician Al-Khwarizmi, whose work on algebra was translated into Arabic and became influential in the development of mathematical terminology.

Arabic scientific terminology also reflects the influence of Islamic theology and philosophy. Many scientific terms in Arabic are derived from Quranic and Islamic theological concepts, reflecting the close relationship between science and religion in Islamic civilization. For example, the Arabic word for "science" (ilm) is derived from the Quranic term for knowledge, emphasizing the importance of knowledge and learning in Islam.

One of the most important aspects of Arabic scientific terminology is its adaptability and flexibility. Arabic scholars were not bound by strict linguistic rules and were able to create new terms and concepts as needed to describe new scientific discoveries. This flexibility allowed Arabic scientific terminology to evolve and grow over time, reflecting the dynamic nature of scientific knowledge.

In conclusion, the Arabic language played a crucial role in the development and dissemination of scientific knowledge in Islamic civilization. Arabic scientific terminology was a key tool that helped Muslim scholars and scientists communicate and share knowledge across cultures and generations. The rich and precise scientific terminology developed in Arabic continues to be used in scientific discourse today, highlighting the enduring legacy of Arabic in the field of science and learning.

Chapter 25: Islamic Art and Its Influence on Science

Islamic art, with its intricate geometric patterns, arabesques, and calligraphy, has had a profound influence on Islamic science, particularly in the fields of mathematics, astronomy, and optics. Islamic art is not just decorative; it reflects a deep understanding of mathematical principles and scientific concepts that were integral to Islamic civilization.

One of the key ways in which Islamic art influenced science was through its use of geometric patterns. Islamic artists used a variety of geometric shapes, such as circles, squares, and polygons, to create intricate designs that were both aesthetically pleasing and mathematically precise. These geometric patterns were not just decorative; they were also symbolic, representing the underlying order and harmony of the universe as understood in Islamic cosmology.

Islamic artists also used arabesques, or flowing vegetal motifs, in their designs. Arabesques are characterized by their rhythmic repetition and intricate interlacing, which create a sense of infinite expansion and unity. Arabesques were not just decorative motifs; they were also symbolic of the interconnectedness of all things in the universe, reflecting the Islamic belief in the unity of God and creation.

Calligraphy, the art of beautiful writing, was another important aspect of Islamic art. Islamic calligraphy was not just a form of artistic expression; it was also a way of preserving and transmitting knowledge. Many important scientific texts were written in beautiful calligraphy, which helped to elevate the status of these texts and make them more accessible to a wider audience.

Islamic art also influenced the development of scientific instruments. Muslim craftsmen and artisans were responsible for creating many of the instruments used by Muslim astronomers and mathematicians, such as astrolabes and quadrants. These instruments were not just functional; they were also works of art, with intricate designs and decorative elements that reflected the aesthetic sensibilities of Islamic art.

In conclusion, Islamic art had a profound influence on Islamic science, particularly in the fields of mathematics, astronomy, and optics. Islamic artists used geometric patterns, arabesques, and calligraphy to create works of art that were both beautiful and scientifically sophisticated. The influence of Islamic art on science is a testament to the interconnectedness of art and science in Islamic civilization, where beauty and knowledge were seen as two sides of the same coin.

Chapter 26: Astronomy and Timekeeping in Islamic Civilization

Astronomy played a central role in Islamic civilization, with Muslim astronomers making significant advancements in the field and contributing to the development of new theories and technologies. Astronomy was not only pursued for its scientific value but also for its practical applications, particularly in the fields of timekeeping, navigation, and religious observance.

One of the key contributions of Islamic astronomers was the development of sophisticated instruments for measuring time and tracking celestial movements. Muslim astronomers invented a variety of instruments, such as the astrolabe, the quadrant, and the armillary sphere, which were used for making accurate astronomical observations and calculations. These instruments were crucial for determining the direction of Mecca for prayers (qibla), calculating the times for daily prayers (salat), and establishing the beginning of the lunar months for religious festivals.

Islamic astronomers also made significant advancements in the field of celestial mechanics, developing models to explain the movements of celestial bodies. One of the most influential Muslim astronomers was Al-Battani (c. 858–929 CE), who refined the existing Ptolemaic model of the solar system and made more accurate measurements of the length of the year and the tilt of the Earth's axis.

Another important Muslim astronomer was Al-Zarqali (Azarquiel) (1029–1087 CE), who developed a new astronomical instrument called the "zarqal," which was used for measuring the positions of celestial objects. Al-Zarqali's work on the motion of the stars and planets was highly influential and helped to advance the field of astronomy in Islamic civilization.

Islamic astronomers also made important contributions to the field of timekeeping. Muslim scholars developed sophisticated methods for measuring time, including the division of the day into 24 hours, the use of sundials and water clocks, and the development of accurate calendars. The Islamic calendar, based on the lunar year, was particularly important for religious observance and for determining the dates of religious festivals.

In conclusion, astronomy and timekeeping played a central role in Islamic civilization, with Muslim astronomers making significant advancements in the field and contributing to the development of new theories and technologies. Astronomy was not only pursued for its scientific value but also for its practical applications, particularly in the fields of timekeeping, navigation, and religious observance. The contributions of Islamic astronomers continue to be recognized today as important milestones in the history of science and astronomy.

Chapter 27: Islamic Contributions to Psychology and Psychiatry

Islamic civilization made significant contributions to the field of psychology and psychiatry, with Muslim scholars and physicians developing innovative theories and treatments for mental health disorders. Islamic psychology was deeply influenced by Islamic theology and philosophy, and it emphasized the importance of the mind, body, and spirit in maintaining mental health.

One of the key concepts in Islamic psychology is the concept of the "nafs," which refers to the human soul or psyche. Islamic scholars developed a sophisticated understanding of the nafs, categorizing it into different levels or stages, such as the nafs al-ammara (the commanding self), the nafs al-lawwama (the self-reproaching self), and the nafs al-mutma'inna (the tranquil self). This concept of the nafs helped to explain the different aspects of human behavior and personality and laid the foundation for later developments in psychology.

Islamic scholars also developed theories about the causes of mental health disorders and proposed treatments based on Islamic principles. For example, Islamic physicians such as Al-Razi (Rhazes) (865–925 CE) and Ibn Sina (Avicenna) (980–1037 CE) wrote extensively about mental health disorders and proposed treatments that included dietary changes, physical exercise, and spiritual practices such as prayer and meditation.

One of the most important contributions of Islamic psychology to the field of psychiatry was the development of the first psychiatric hospitals. Muslim physicians established hospitals in Baghdad, Cairo, and Damascus that specialized in the treatment of mental health disorders. These hospitals provided humane and compassionate care to patients and were staffed by trained physicians and nurses who followed ethical guidelines for the treatment of mental illness.

Islamic scholars also made important advancements in the field of dream interpretation, which was considered a valuable tool for understanding the unconscious mind. Muslim scholars developed elaborate systems for interpreting dreams based on Islamic theology and symbolism, which influenced later developments in the field of psychoanalysis.

In conclusion, Islamic civilization made significant contributions to the field of psychology and psychiatry, with Muslim scholars and physicians developing innovative theories and treatments for mental health disorders. Islamic psychology emphasized the importance of the mind, body, and spirit in maintaining mental health and laid the foundation for later developments in the field of psychology. The contributions of Islamic scholars continue to be recognized today as important milestones in the history of psychology and psychiatry.

Chapter 28: The Legacy of Al-Andalus in Science

Al-Andalus, the Islamic civilization in medieval Spain, had a profound and lasting impact on the development of science and scholarship in Europe and the Islamic world. Al-Andalus was a center of learning and innovation, where Muslim, Jewish, and Christian scholars worked together to advance knowledge in a wide range of fields, including astronomy, mathematics, medicine, and philosophy.

One of the key contributions of Al-Andalus to science was its role in the transmission of Greek, Persian, and Indian knowledge to Europe. Muslim scholars in Al-Andalus translated many important works from these languages into Arabic, preserving them for future generations and making them accessible to scholars in Europe. These translations played a crucial role in the revival of learning in Europe during the Renaissance.

Al-Andalus was also a center of scientific innovation, where Muslim scholars made significant advancements in various fields. One of the most important figures in Andalusian science was Ibn al-Haytham (Alhazen) (965–1040 CE), who made groundbreaking contributions to the field of optics and visual perception. Ibn al-Haytham's work on optics laid the foundation for modern theories of vision and influenced later developments in the field of physics.

Another important Andalusian scientist was Al-Zarqali (Azarquiel) (1029–1087 CE), who made significant advancements in astronomy and developed new instruments for measuring celestial movements. Al-Zarqali's work on the motion of the stars and planets was highly influential and helped to advance the field of astronomy in Europe.

In addition to their scientific achievements, the scholars of Al-Andalus also made important contributions to the field of medicine. Muslim physicians in Al-Andalus, such as Ibn Zuhr (Avenzoar) (1091–1161 CE), made significant advancements in the field of surgery and pharmacology, which helped to advance medical knowledge in Europe.

The legacy of Al-Andalus in science is not just about the individual achievements of its scholars, but also about the spirit of inquiry, tolerance, and intellectual exchange that characterized the civilization. Al-Andalus was a melting pot of cultures and ideas, where scholars from different backgrounds and traditions worked together to advance knowledge and understanding. This spirit of collaboration and openness to new ideas was a hallmark of Al-Andalus and contributed to its status as a center of learning and innovation.

In conclusion, the legacy of Al-Andalus in science is a testament to the richness and diversity of Islamic civilization. The scholars of Al-Andalus made significant contributions to the fields of astronomy, mathematics, medicine, and philosophy, and their work continues to inspire scholars and researchers today. Al-Andalus's legacy reminds us of the importance of intellectual exchange and collaboration in the advancement of knowledge and understanding.

Chapter 29: Islamic Influence on Modern Medicine

Islamic civilization made significant contributions to the field of medicine, which have had a lasting impact on modern medical practice. Islamic physicians and scholars made important advancements in various aspects of medicine, including anatomy, surgery, pharmacology, and public health, and their work continues to influence modern medical practice.

One of the key contributions of Islamic medicine to modern medicine was in the field of anatomy. Muslim physicians such as Ibn al-Nafis (1213–1288 CE) made important discoveries in anatomy, including the circulation of blood through the lungs. Ibn al-Nafis's work on the pulmonary circulation predated similar discoveries in Europe by several centuries and laid the foundation for later developments in the field of cardiology.

Islamic medicine also made significant advancements in the field of surgery. Muslim physicians such as Al-Zahrawi (Albucasis) (936–1013 CE) wrote extensively on surgical techniques and instruments, and their work was highly influential in the development of modern surgical practices. Al-Zahrawi's surgical treatise, "Al-Tasrif," was used as a standard textbook in European medical schools for centuries.

Islamic medicine also made important contributions to pharmacology. Muslim physicians compiled extensive pharmacopoeias that documented the medicinal properties of various plants and herbs. These pharmacopoeias were used as reference guides by physicians and pharmacists and helped to advance the field of herbal medicine.

Islamic medicine also emphasized the importance of public health and hygiene. Muslim physicians developed guidelines for maintaining health and preventing disease, including recommendations for personal hygiene, sanitation, and quarantine. These guidelines were based on Islamic teachings and helped to improve public health in Islamic cities and beyond.

In addition to their specific contributions to medical knowledge, Islamic physicians and scholars also played a crucial role in preserving and transmitting the medical knowledge of ancient civilizations, such as Greece, Rome, Persia, and India. Muslim scholars translated many important medical works from these languages into Arabic, preserving them for future generations and making them accessible to scholars in Europe during the Renaissance.

In conclusion, Islamic civilization made significant contributions to the field of medicine, which have had a lasting impact on modern medical practice. Islamic physicians and scholars made important advancements in anatomy, surgery, pharmacology, and public health, and their work continues to influence modern medical practice. The contributions of Islamic medicine are a testament to the richness and diversity of Islamic civilization and its enduring legacy in the field of medicine.

Chapter 30: The Decline of Islamic Science

The decline of Islamic science refers to a period in Islamic civilization when scientific advancements and scholarly pursuits stagnated or declined. This decline is generally attributed to a combination of political, social, and intellectual factors that affected the Islamic world from the 13th century onwards.

One of the key factors contributing to the decline of Islamic science was the political fragmentation and instability that characterized the Islamic world during this period. The Abbasid Caliphate, which had been a center of learning and scholarship, began to decline in the 9th century, leading to the fragmentation of the Islamic world into smaller, often warring, states. This fragmentation disrupted the flow of knowledge and hindered the development of scientific institutions and research.

Another factor contributing to the decline of Islamic science was the decline of patronage for learning and scholarship. As the political and economic fortunes of the Islamic world declined, so too did the patronage of scholars and scientists. Many scholars were forced to seek patronage from local rulers or wealthy individuals, which often resulted in a lack of funding and support for scientific research.

The decline of Islamic science was also influenced by the rise of anti-intellectual and anti-rationalist movements within Islamic society. Some scholars and religious leaders began to reject the study of natural sciences and philosophy, arguing that they were contrary to Islamic teachings. This anti-intellectualism stifled scientific inquiry and hindered the development of new ideas and theories.

The decline of Islamic science was also affected by external factors, such as the Mongol invasions of the 13th and 14th centuries. The Mongol invasions devastated many parts of the Islamic world, destroying libraries, universities, and scientific institutions, and disrupting the flow of knowledge.

Despite these challenges, it is important to note that the decline of Islamic science was not a complete or irreversible process. Islamic scholars and scientists continued to make important contributions to various fields of knowledge, and Islamic civilization continued to be a center of learning and scholarship. The decline of Islamic science was a complex and multifaceted process, and its causes and effects continue to be debated by scholars today.

Chapter 31: Revival of Interest in Islamic Science

The revival of interest in Islamic science refers to a renewed appreciation and study of the scientific achievements of Islamic civilization, particularly during the medieval period. This revival began in the 19th century and continues to the present day, as scholars and researchers seek to rediscover and reinterpret the scientific legacy of Islamic civilization.

One of the key factors contributing to the revival of interest in Islamic science was the translation of important scientific works from Arabic into European languages. In the 19th and early 20th centuries, European scholars began to translate many important scientific texts from Arabic, Persian, and other Islamic languages into English, French, and German. These translations helped to reintroduce Islamic scientific ideas and discoveries to a wider audience and stimulated new research in the field.

The revival of interest in Islamic science was also influenced by broader intellectual and cultural movements, such as Orientalism and the Romantic movement. Orientalist scholars, such as Edward Said, emphasized the importance of Islamic civilization in the development of world history and culture, sparking a renewed interest in Islamic science and scholarship. The Romantic movement, with its emphasis on the exotic and the mystical, also contributed to a romanticized view of Islamic science and its achievements.

Another factor contributing to the revival of interest in Islamic science was the recognition of the scientific achievements of Islamic civilization by modern historians and scholars. Scholars such as George Sarton and Bernard Lewis highlighted the contributions of Islamic scholars and scientists to various fields of knowledge, including mathematics, astronomy, medicine, and philosophy. This recognition helped to raise awareness of the importance of Islamic science in the history of science and technology.

The revival of interest in Islamic science has also been driven by the desire to understand the roots of modern science and technology. Many modern scientific concepts and technologies have their origins in Islamic science, and scholars and researchers have sought to trace the development of these ideas from their Islamic origins to their modern manifestations. This research has helped to highlight the enduring legacy of Islamic science and its impact on the modern world.

In conclusion, the revival of interest in Islamic science has been driven by a variety of factors, including the translation of important texts, intellectual and cultural movements, and the recognition of Islamic scientific achievements by modern scholars. This revival has helped to shed light on the rich scientific heritage of Islamic civilization and its continuing relevance to the modern world.

Chapter 32: Modern Interpretations of Islamic Science

Modern interpretations of Islamic science refer to the ways in which scholars and researchers today understand and interpret the scientific achievements of Islamic civilization, particularly during the medieval period. These interpretations are informed by a variety of factors, including historical research, cultural perspectives, and contemporary debates in science and religion.

One of the key aspects of modern interpretations of Islamic science is a recognition of the diversity and complexity of Islamic scientific thought. Islamic scholars and scientists came from a variety of cultural and religious backgrounds, and their work was influenced by a wide range of intellectual traditions, including Greek, Persian, Indian, and Chinese. Modern scholars seek to understand this diversity and to appreciate the contributions of scholars from different backgrounds and traditions.

Modern interpretations of Islamic science also emphasize the interdisciplinary nature of Islamic scientific thought. Islamic scholars and scientists did not compartmentalize knowledge into separate disciplines, as is often done in modern academia. Instead, they saw knowledge as interconnected and holistic, with different branches of knowledge informing and enriching each other. Modern scholars seek to understand this holistic approach to knowledge and to appreciate its relevance to contemporary scholarship.

Another key aspect of modern interpretations of Islamic science is a reassessment of the relationship between science and religion in Islamic civilization. While it is often assumed that science and religion were in conflict in Islamic civilization, many modern scholars argue that this was not always the case. Islamic scholars and scientists saw science as a way of understanding the natural world and as a means of deepening their faith in God. Modern scholars seek to understand this harmonious relationship between science and religion and to appreciate its implications for contemporary science and religion debates.

Modern interpretations of Islamic science also seek to highlight the relevance of Islamic scientific thought to contemporary issues and challenges. Islamic scholars and scientists made significant advancements in fields such as medicine, astronomy, and mathematics, which continue to be relevant to modern scientific research. Modern scholars seek to build on this legacy and to apply Islamic scientific principles to contemporary scientific and technological developments.

In conclusion, modern interpretations of Islamic science are informed by a variety of factors, including historical research, cultural perspectives, and contemporary debates in science and religion. These interpretations seek to appreciate the diversity and complexity of Islamic scientific thought, to understand its interdisciplinary nature, and to reassess its relationship to religion. By engaging with the scientific achievements of Islamic civilization, modern scholars seek to enrich our understanding of both the past and the present.

Chapter 33: Islamic Ethics and Biotechnology

Islamic ethics play a significant role in shaping attitudes towards biotechnology, particularly in areas such as genetic engineering, stem cell research, and reproductive technologies. Islamic scholars and ethicists have engaged with these issues, drawing on Islamic principles and values to guide ethical decision-making in biotechnology.

One of the key principles of Islamic ethics relevant to biotechnology is the concept of "maslaha," or the greater good. Islamic scholars consider the potential benefits and harms of biotechnologies, weighing them against each other to determine whether a particular technology is permissible (halal) or prohibited (haram). This principle emphasizes the importance of considering the wider social and ethical implications of biotechnologies, rather than just their immediate benefits or harms.

Islamic ethics also emphasize the sanctity of life and the protection of human dignity. This principle is particularly relevant to issues such as genetic engineering and stem cell research, where there is potential for manipulation of the human genome or destruction of embryos. Islamic scholars and ethicists have debated these issues, with some arguing that such technologies are permissible if they are used for legitimate medical purposes and do not violate the sanctity of life.

Another key principle of Islamic ethics is the concept of "adl," or justice. Islamic scholars emphasize the importance of ensuring that biotechnologies are used in a fair and equitable manner, without causing harm or injustice to individuals or society as a whole. This principle is particularly relevant to issues such as access to healthcare and the distribution of benefits and risks associated with biotechnologies.

Islamic ethics also emphasize the importance of respecting natural boundaries and limits. Islamic scholars argue that biotechnologies should not be used to exceed these limits or to interfere with the natural order of creation. This principle is particularly relevant to issues such as cloning and genetic modification, where there is potential for the manipulation of natural processes.

In conclusion, Islamic ethics play a significant role in shaping attitudes towards biotechnology, guiding ethical decision-making in areas such as genetic engineering, stem cell research, and reproductive technologies. Islamic scholars and ethicists draw on Islamic principles and values to determine whether a particular biotechnology is permissible or prohibited, considering factors such as the greater good, the sanctity of life, justice, and respect for natural boundaries. By engaging with these ethical principles, Islamic scholars and ethicists seek to ensure that biotechnologies are used in a manner that is consistent with Islamic values and principles.

Chapter 34: Islamic Science and Environmental Conservation

Islamic science has a long history of emphasizing the importance of environmental conservation and sustainability. Islamic teachings and principles provide a framework for understanding humanity's relationship with the natural world and offer guidance on how to protect and preserve the environment for future generations.

One of the key principles of Islamic environmental ethics is the concept of "Khalifah," or stewardship. According to Islamic teachings, humans are considered stewards of the Earth, responsible for managing its resources wisely and protecting its natural balance. This concept emphasizes the importance of sustainable practices that do not deplete or harm the environment.

Islamic teachings also emphasize the importance of balance and moderation in all aspects of life, including in the use of natural resources. The Quran instructs believers to "eat and drink, but waste not by excess, for Allah loves not the wasters" (Quran 7:31), highlighting the importance of avoiding waste and extravagance in the use of resources.

Islamic scholars and ethicists have also drawn on the concept of "Amanah," or trust, to emphasize the importance of protecting the environment. Muslims are taught to consider themselves as trustees of the Earth, responsible for preserving its beauty and diversity for future generations.

Islamic science has also contributed to environmental conservation through its emphasis on observation and understanding of the natural world. Muslim scientists and scholars made significant advancements in fields such as botany, zoology, and ecology, documenting the diversity of plant and animal life and recognizing the interconnectedness of all living beings.

In modern times, Islamic scholars and organizations have been actively involved in promoting environmental conservation and sustainability. Islamic environmental organizations, such as the Islamic Foundation for Ecology and Environmental Sciences (IFEES), work to raise awareness about environmental issues and promote sustainable practices based on Islamic principles.

In conclusion, Islamic science has a rich tradition of emphasizing the importance of environmental conservation and sustainability. Islamic teachings and principles provide a framework for understanding humanity's role as stewards of the Earth and offer guidance on how to protect and preserve the environment for future generations. By drawing on these teachings, Muslims can contribute to global efforts to address environmental challenges and build a more sustainable future.

Chapter 35: Islamic Mathematics and Modern Cryptography

Islamic mathematics has had a significant influence on the development of modern cryptography, the science of secure communication. Cryptography involves creating codes and ciphers to protect sensitive information from unauthorized access. Islamic mathematicians made important contributions to cryptography through their work on number theory, algebra, and encryption methods.

One of the key contributions of Islamic mathematics to cryptography was the development of sophisticated encryption methods. Muslim mathematicians, such as Al-Kindi (801–873 CE), developed methods for encrypting messages using substitution ciphers, where letters are replaced with other letters or symbols according to a specific algorithm. These methods were used to encode sensitive information and protect it from being intercepted by unauthorized parties.

Islamic mathematics also made important contributions to number theory, which is the branch of mathematics that deals with the properties of numbers. Muslim mathematicians, such as Al-Khwarizmi (780–850 CE), made significant advancements in number theory, including the development of algebra and the use of mathematical symbols to represent unknown quantities. These advancements laid the foundation for modern encryption techniques, which rely on complex mathematical algorithms to secure data.

The influence of Islamic mathematics on modern cryptography can be seen in the use of algorithms and mathematical principles derived from Islamic scholarship. For example, the RSA algorithm, which is widely used in modern encryption systems, is based on the principles of number theory developed by Islamic mathematicians. The RSA algorithm uses prime numbers to create encryption keys, which are then used to encrypt and decrypt messages securely.

In addition to their contributions to encryption methods, Islamic mathematicians also made important advancements in the field of cryptanalysis, which is the science of breaking codes and ciphers. Muslim scholars developed methods for breaking encryption codes and deciphering encrypted messages, which were used to protect sensitive information and gain intelligence advantage in military and diplomatic communications.

In conclusion, Islamic mathematics has had a profound influence on the development of modern cryptography. Muslim mathematicians made important contributions to encryption methods, number theory, and cryptanalysis, which have helped to shape the field of secure communication. By building on the mathematical principles developed by Islamic scholars, modern cryptographers have been able to create sophisticated encryption systems that protect sensitive information in today's digital age.

Chapter 36: Islamic Medicine and Modern Healthcare

Islamic medicine has had a significant impact on the development of modern healthcare, with many principles and practices from Islamic medicine influencing contemporary medical practice. Islamic medicine was characterized by a holistic approach to health, with a focus on preventive care, ethical considerations, and the integration of spiritual and physical well-being.

One of the key principles of Islamic medicine that has influenced modern healthcare is the emphasis on preventive care. Islamic medicine emphasizes the importance of maintaining a healthy lifestyle, including proper diet, exercise, and hygiene, to prevent illness and promote well-being. This focus on preventive care is reflected in modern healthcare practices, which prioritize wellness and disease prevention.

Islamic medicine also emphasizes the importance of ethical considerations in healthcare. Islamic scholars developed guidelines for medical ethics, emphasizing the importance of compassion, honesty, and integrity in the practice of medicine. These ethical principles have influenced modern healthcare practices, shaping the way healthcare professionals interact with patients and make decisions about their care.

Another key aspect of Islamic medicine that has influenced modern healthcare is the integration of spiritual and physical well-being. Islamic medicine views health as a balance of body, mind, and spirit, and emphasizes the importance of spiritual practices such as prayer and meditation in maintaining health. This holistic approach to health has influenced modern healthcare practices, with many healthcare providers recognizing the importance of addressing the spiritual and emotional needs of patients in addition to their physical health.

Islamic medicine has also made important contributions to the field of pharmacology. Muslim scholars developed extensive pharmacopoeias that documented the medicinal properties of plants and herbs, many of which are still used in modern medicine. Islamic medicine also made advancements in the field of surgery, developing new techniques and instruments that laid the foundation for modern surgical practices.

In conclusion, Islamic medicine has had a significant impact on the development of modern healthcare. Its emphasis on preventive care, ethical considerations, and holistic well-being has influenced contemporary medical practice, shaping the way healthcare is delivered and the principles that guide it. By drawing on the principles and practices of Islamic medicine, modern healthcare providers can continue to improve the quality of care and promote well-being for all.

Chapter 37: Islamic Architecture and Modern Engineering

Islamic architecture has had a profound influence on modern engineering, with many principles and techniques from Islamic architecture being used in contemporary building design and construction. Islamic architecture is characterized by its emphasis on geometric patterns, decorative elements, and innovative structural systems, all of which have influenced modern engineering practices.

One of the key aspects of Islamic architecture that has influenced modern engineering is the use of geometric patterns and decorative elements. Islamic architecture is known for its intricate geometric patterns, which are often used to decorate surfaces such as walls, ceilings, and floors. These patterns are not only aesthetically pleasing but also serve a structural purpose, helping to distribute weight and provide stability to the building.

Islamic architecture is also known for its innovative structural systems, such as the use of domes, arches, and vaults. These structural elements are not only visually striking but also serve functional purposes, such as providing support for large open spaces and allowing for the use of natural light and ventilation. Modern engineers have drawn on these techniques in the design of modern buildings, using similar structural systems to create efficient and aesthetically pleasing structures.

Another aspect of Islamic architecture that has influenced modern engineering is the use of materials. Islamic architects were skilled in working with a variety of materials, including stone, brick, wood, and ceramics, and they developed innovative techniques for using these materials in their buildings. Modern engineers have drawn on these techniques, using a variety of materials in their designs and developing new techniques for working with them.

Islamic architecture is also known for its use of water features, such as fountains, pools, and channels, which are used for both decorative and practical purposes. These water features not only add to the beauty of the building but also serve functional purposes, such as providing cooling and humidity control. Modern engineers have incorporated water features into their designs, using them to enhance the aesthetic appeal and functionality of modern buildings.

In conclusion, Islamic architecture has had a significant influence on modern engineering, with many principles and techniques from Islamic architecture being used in contemporary building design and construction. By drawing on the principles and techniques of Islamic architecture, modern engineers can create buildings that are not only functional and efficient but also beautiful and culturally significant.

Chapter 38: Islamic Astronomy and Modern Space Exploration

Islamic astronomy has played a significant role in the development of modern space exploration, with many concepts and techniques from Islamic astronomy influencing the field of astronomy and space science. Islamic astronomers made important contributions to the study of the stars, planets, and celestial phenomena, laying the foundation for modern astronomy and space exploration.

One of the key contributions of Islamic astronomy to modern space exploration is the development of observational techniques and instruments. Islamic astronomers developed sophisticated instruments, such as the astrolabe and the quadrant, for measuring the positions of celestial objects. These instruments were used to make accurate observations of the stars and planets, providing valuable data for astronomers and navigators.

Islamic astronomy also made important advancements in the field of celestial mechanics, the study of the motion of celestial bodies. Muslim astronomers developed models to explain the movements of the stars and planets, including the development of the heliocentric model of the solar system. These advancements laid the foundation for modern theories of planetary motion and space travel.

The influence of Islamic astronomy on modern space exploration can also be seen in the naming of stars and celestial objects. Many stars and constellations have names of Arabic origin, reflecting the contributions of Islamic astronomers to the study of the night sky. For example, the star Aldebaran is derived from the Arabic word "al-dabaran," which means "the follower," referring to its position behind the Pleiades star cluster.

Islamic astronomy also influenced the development of calendars, which are used to track the movements of the sun, moon, and stars. The Islamic calendar, based on the lunar year, is used to determine the dates of religious festivals and observances. The Islamic calendar has also been used by astronomers to track the movement of the moon and to calculate the timing of celestial events.

In conclusion, Islamic astronomy has had a significant impact on modern space exploration, with many concepts and techniques from Islamic astronomy influencing the field of astronomy and space science. By building on the contributions of Islamic astronomers, modern scientists and engineers have been able to develop new technologies and techniques for exploring the universe and understanding our place in it.

Chapter 39: Islamic Science and Globalization

Islamic science has played a significant role in the process of globalization, influencing the exchange of knowledge, ideas, and technologies between different cultures and civilizations. Islamic science has been a bridge between East and West, contributing to the development of science and technology in both Islamic and non-Islamic societies.

One of the ways in which Islamic science has contributed to globalization is through the transmission of knowledge. Islamic scholars and scientists preserved and translated ancient Greek, Persian, Indian, and Chinese texts into Arabic, making them accessible to scholars in Europe and other parts of the world. This transmission of knowledge helped to spark a revival of learning in Europe during the Renaissance and contributed to the development of modern science and technology.

Islamic science has also contributed to globalization through its emphasis on collaboration and exchange. Islamic scholars and scientists worked closely with scholars from different cultures and backgrounds, sharing ideas, techniques, and discoveries. This spirit of collaboration helped to advance scientific knowledge and understanding, leading to new discoveries and innovations.

Another way in which Islamic science has contributed to globalization is through its influence on art, architecture, and culture. Islamic architecture, with its intricate geometric patterns and innovative structural techniques, has inspired architects and designers around the world. Islamic art, with its beautiful calligraphy and intricate designs, has influenced artists and designers in diverse cultures.

Islamic science has also contributed to globalization through its influence on medicine, mathematics, astronomy, and other fields. Muslim scholars and scientists made important advancements in these fields, which have had a lasting impact on modern science and technology. For example, the work of Islamic astronomers laid the foundation for modern astronomy, while the advancements made in medicine have helped to shape modern healthcare practices.

In conclusion, Islamic science has played a significant role in the process of globalization, influencing the exchange of knowledge, ideas, and technologies between different cultures and civilizations. By contributing to the development of science, technology, art, and culture, Islamic science has helped to create a more interconnected and diverse world.

Chapter 40: The Future of Islamic Science

The future of Islamic science holds great promise, with opportunities for innovation, collaboration, and advancement in various fields of knowledge. Islamic science has a rich tradition of inquiry and discovery, and its principles and practices continue to inspire scholars and scientists around the world.

One of the key areas of future growth in Islamic science is in the field of technology. Islamic countries are increasingly investing in research and development in areas such as information technology, biotechnology, and renewable energy. These investments are helping to drive innovation and economic growth in Islamic countries, and are contributing to the global scientific community.

Another area of future growth in Islamic science is in the field of environmental science. Islamic teachings emphasize the importance of environmental conservation and sustainability, and Islamic scholars and scientists are increasingly focusing on ways to address environmental challenges such as climate change, deforestation, and pollution. By drawing on Islamic principles and practices, researchers in this field are developing innovative solutions to environmental problems that can benefit both Islamic and non-Islamic societies.

Islamic science also has the potential to contribute to the fields of medicine and healthcare. Islamic scholars and scientists have a long history of advancements in medicine, and contemporary researchers are continuing this tradition by developing new treatments and technologies. Islamic countries are also investing in healthcare infrastructure and education, which are helping to improve healthcare outcomes for millions of people.

In addition to these areas, the future of Islamic science is likely to be shaped by advancements in education and research. Islamic countries are increasingly investing in education and research infrastructure, and are working to attract and retain top talent in scientific fields. By investing in education and research, Islamic countries are laying the foundation for future advancements in science and technology.

Overall, the future of Islamic science is bright, with opportunities for growth and advancement in a wide range of fields. By building on its rich tradition of inquiry and discovery, Islamic science has the potential to make significant contributions to the global scientific community and to address some of the most pressing challenges facing humanity.

Chapter 41: Challenges and Opportunities in Islamic Science

Islamic science faces a range of challenges and opportunities in the modern world. These challenges include issues related to funding, infrastructure, education, and the integration of Islamic principles with modern scientific practices. However, there are also many opportunities for Islamic science to thrive and make valuable contributions to the global scientific community.

One of the key challenges facing Islamic science is the issue of funding. Many Islamic countries struggle to allocate sufficient funds for scientific research and development, which can hinder the ability of researchers to conduct high-quality research and compete on the global stage. Addressing this challenge will require increased investment in science and technology, as well as the development of partnerships with international organizations and private sector entities.

Infrastructure is another challenge facing Islamic science, particularly in terms of research facilities and equipment. Many Islamic countries lack the necessary infrastructure to support cutting-edge scientific research, which can limit the ability of researchers to conduct meaningful research. Addressing this challenge will require investment in research infrastructure, as well as the development of collaborations with institutions in other countries.

Education is also a significant challenge in Islamic science, with many countries lacking the necessary educational resources and expertise to train the next generation of scientists. Improving science education will require investment in teacher training, curriculum development, and the provision of educational materials and resources.

Another challenge facing Islamic science is the integration of Islamic principles with modern scientific practices. While Islamic science has a rich tradition of inquiry and discovery, there are sometimes tensions between Islamic teachings and modern scientific theories. Addressing this challenge will require a nuanced approach that respects both Islamic principles and the principles of modern science.

Despite these challenges, there are also many opportunities for Islamic science to thrive and make valuable contributions to the global scientific community. One of the key opportunities lies in the field of technology, where Islamic countries are increasingly investing in research and development. By harnessing the potential of technology, Islamic science can drive innovation and economic growth in Islamic countries.

Another opportunity for Islamic science lies in the field of environmental science, where Islamic teachings on environmental conservation and sustainability can inform and inspire scientific research. By drawing on Islamic principles and practices, researchers in this field can develop innovative solutions to environmental challenges that benefit both Islamic and non-Islamic societies.

In conclusion, Islamic science faces a range of challenges, but also many opportunities for growth and advancement. By addressing these challenges and seizing these opportunities, Islamic science can make valuable contributions to the global scientific community and help to address some of the most pressing challenges facing humanity.

Chapter 42: The Role of Islamic Science in Interfaith Dialogue

Islamic science can play a significant role in interfaith dialogue by promoting mutual understanding, respect, and collaboration among people of different religious backgrounds. Islamic science has a rich tradition of inquiry and discovery that can serve as a common ground for dialogue and exchange between different faith communities.

One of the key ways in which Islamic science can contribute to interfaith dialogue is by highlighting the shared values and principles of different religions. Islamic teachings emphasize the importance of seeking knowledge, understanding the natural world, and caring for the environment, principles that are shared by many other religious traditions. By emphasizing these shared values, Islamic science can help to foster a sense of common purpose and cooperation among people of different faiths.

Islamic science can also contribute to interfaith dialogue by promoting a more holistic and integrated approach to knowledge. Islamic scholars and scientists historically viewed knowledge as interconnected and holistic, with different branches of knowledge informing and enriching each other. This holistic approach to knowledge can help to bridge the gap between different fields of study and promote a more integrated understanding of the world.

Another way in which Islamic science can contribute to interfaith dialogue is by fostering a spirit of curiosity and open-mindedness. Islamic scholars and scientists historically embraced a spirit of inquiry and curiosity, seeking to understand the natural world and the mysteries of the universe. This spirit of curiosity can help to break down barriers between different faith communities and promote a more open and inclusive approach to dialogue.

Islamic science can also contribute to interfaith dialogue by promoting ethical and responsible use of technology. Islamic teachings emphasize the importance of ethical behavior and responsible stewardship of the Earth, principles that are increasingly relevant in the context of modern technological advancements. By promoting ethical use of technology, Islamic science can help to address common concerns and challenges faced by people of different faiths.

In conclusion, Islamic science has the potential to play a significant role in interfaith dialogue by promoting mutual understanding, respect, and collaboration among people of different religious backgrounds. By highlighting shared values, promoting a holistic approach to knowledge, fostering curiosity and open-mindedness, and promoting ethical use of technology, Islamic science can help to build bridges between different faith communities and promote a more peaceful and harmonious world.

Chapter 43: Islamic Science and Ethics in Artificial Intelligence

Islamic science offers a unique perspective on the ethical considerations surrounding artificial intelligence (AI), emphasizing the importance of ethical behavior, justice, and compassion in the development and use of AI technologies. Islamic scholars and ethicists have engaged with the ethical implications of AI, drawing on Islamic principles and values to guide ethical decision-making in this rapidly advancing field.

One of the key ethical considerations in AI is the impact of AI technologies on society and the environment. Islamic teachings emphasize the importance of considering the broader societal and environmental impacts of technology, and Islamic ethicists have called for AI technologies to be developed and used in a way that promotes the well-being of all individuals and communities.

Islamic ethics also emphasizes the importance of justice and fairness in the development and use of AI technologies. Islamic scholars and ethicists have raised concerns about the potential for AI technologies to exacerbate existing inequalities and discrimination, and have called for safeguards to ensure that AI technologies are used in a fair and equitable manner.

Another key ethical consideration in AI is the issue of autonomy and human dignity. Islamic teachings emphasize the importance of respecting human dignity and autonomy, and Islamic ethicists have raised concerns about the potential for AI technologies to infringe on these principles.

Islamic scholars and ethicists have called for the development of AI technologies that respect human autonomy and dignity, and that are used to enhance human capabilities rather than replace them.

Islamic ethics also emphasizes the importance of compassion and empathy in human interactions. Islamic scholars and ethicists have called for AI technologies to be developed and used in a way that promotes compassion and empathy, and that enhances human relationships rather than diminishes them.

In conclusion, Islamic science offers valuable insights into the ethical considerations surrounding artificial intelligence. By drawing on Islamic principles and values, researchers and developers can develop AI technologies that are ethical, just, and compassionate, and that contribute to the well-being of individuals and communities.

Chapter 44: Islamic Science and Sustainable Development Goals

Islamic science can play a significant role in achieving the Sustainable Development Goals (SDGs) set by the United Nations, which aim to address global challenges such as poverty, hunger, health, education, gender equality, clean water, sustainable energy, economic growth, and climate action. Islamic teachings and principles provide a framework for understanding and addressing these challenges, and Islamic science can help to inform and guide efforts to achieve the SDGs.

One of the key ways in which Islamic science can contribute to the SDGs is through its emphasis on sustainability and environmental conservation. Islamic teachings emphasize the importance of caring for the environment and using natural resources wisely, principles that are central to many of the SDGs related to environmental sustainability. By drawing on Islamic principles and practices, policymakers and practitioners can develop sustainable solutions to environmental challenges that are aligned with the SDGs.

Islamic science can also contribute to the SDGs by promoting social justice and equity. Islamic teachings emphasize the importance of fairness, compassion, and solidarity, principles that are central to many of the SDGs related to poverty, hunger, health, education, and gender equality. By drawing on Islamic principles and values, policymakers and practitioners can develop strategies that promote social justice and equity and help to achieve the SDGs.

Another way in which Islamic science can contribute to the SDGs is through its emphasis on education and knowledge. Islamic teachings emphasize the importance of seeking knowledge and understanding the natural world, principles that are central to many of the SDGs related to education, innovation, and economic growth. By drawing on Islamic principles and practices, policymakers and practitioners can develop educational programs that promote sustainable development and help to achieve the SDGs.

In conclusion, Islamic science can play a significant role in achieving the Sustainable Development Goals by promoting sustainability, social justice, and education. By drawing on Islamic principles and values, policymakers and practitioners can develop strategies that address the root causes of global challenges and help to create a more sustainable and equitable world.

Chapter 45: Islamic Science and Climate Change

Islamic science offers valuable insights and principles that can inform efforts to address climate change, one of the most pressing challenges facing the world today. Islamic teachings emphasize the importance of stewardship of the Earth and the protection of the environment, principles that are central to addressing the causes and impacts of climate change.

One of the key principles of Islamic science that can inform efforts to address climate change is the concept of "Khalifah," or stewardship. According to Islamic teachings, humans are considered stewards of the Earth, responsible for managing its resources wisely and protecting its natural balance. This principle emphasizes the importance of sustainable practices that do not harm the environment or deplete its resources.

Islamic teachings also emphasize the importance of justice and equity, principles that are central to addressing the impacts of climate change on vulnerable communities. Islamic scholars and ethicists have called for action to address climate change, highlighting the disproportionate impact of climate change on the poor and marginalized. By drawing on Islamic principles of justice and equity, policymakers and practitioners can develop strategies that address the impacts of climate change on vulnerable communities and promote a more just and equitable response to the climate crisis.

Islamic science also emphasizes the importance of knowledge and understanding, principles that are central to addressing the complex challenges of climate change. Islamic scholars and scientists have called for increased research and education on climate change, highlighting the importance of understanding the causes and impacts of climate change in order to develop effective solutions.

In conclusion, Islamic science offers valuable insights and principles that can inform efforts to address climate change. By drawing on Islamic teachings of stewardship, justice, equity, and knowledge, policymakers and practitioners can develop strategies that promote sustainable development and help to mitigate the impacts of climate change on vulnerable communities.

Chapter 46: Islamic Science and the Digital Age

Islamic science is experiencing a renaissance in the digital age, with advancements in technology and communication opening up new opportunities for research, collaboration, and innovation. Islamic scholars and scientists are leveraging digital tools and platforms to further their understanding of the natural world, promote ethical and sustainable practices, and engage with global challenges.

One of the key ways in which Islamic science is thriving in the digital age is through the use of digital tools and platforms for research and collaboration. Islamic scholars and scientists are using digital databases, online journals, and collaborative platforms to access and share research findings, collaborate with colleagues around the world, and disseminate their work to a global audience. These digital tools have made it easier for researchers to connect and collaborate across geographical and cultural boundaries, leading to new insights and discoveries in Islamic science.

Islamic science is also benefiting from advancements in digital technology, such as artificial intelligence (AI), machine learning, and big data analytics. These technologies are being used to analyze large datasets, model complex systems, and simulate natural phenomena, leading to new advancements in fields such as astronomy, medicine, and environmental science. Islamic scholars and scientists are also using AI and machine learning to develop innovative solutions to global challenges, such as climate change, poverty, and disease.

In addition to advancements in research and technology, Islamic science is also thriving in the digital age through increased access to education and knowledge. Online learning platforms and digital libraries are making it easier for people around the world to access Islamic teachings and scholarship, leading to a greater understanding and appreciation of Islamic science and its contributions to the world.

Overall, the digital age is providing new opportunities for Islamic science to thrive and make valuable contributions to the global scientific community. By leveraging digital tools and platforms, Islamic scholars and scientists are advancing knowledge, promoting ethical and sustainable practices, and engaging with global challenges in ways that were not possible before.

Chapter 47: Islamic Science and Social Justice

Islamic science has a long history of emphasizing social justice, with Islamic teachings and principles providing a framework for understanding and addressing social inequalities and injustices. Islamic scholars and scientists have developed theories and practices that promote social justice, and Islamic science continues to play a role in advancing social justice in the modern world.

One of the key ways in which Islamic science promotes social justice is through its emphasis on ethical behavior and compassion. Islamic teachings emphasize the importance of treating others with kindness and compassion, and Islamic scholars and scientists have developed ethical guidelines for behavior that promote fairness, equality, and justice. By drawing on these principles, Islamic science can help to address social inequalities and promote a more just and equitable society.

Islamic science also promotes social justice through its emphasis on knowledge and education. Islamic teachings emphasize the importance of seeking knowledge and understanding the world, and Islamic scholars and scientists have developed educational systems that promote the intellectual and moral development of individuals. By promoting education and knowledge, Islamic science can empower individuals to advocate for their rights and address social injustices.

Another way in which Islamic science promotes social justice is through its emphasis on community and solidarity. Islamic teachings emphasize the importance of community and encourage individuals to support and care for one another. By fostering a sense of community and solidarity, Islamic science can help to build networks of support that can address social inequalities and injustices.

In addition to these principles, Islamic science also promotes social justice through its emphasis on economic fairness and sustainability. Islamic teachings emphasize the importance of economic justice and prohibit practices such as usury and exploitation. By promoting economic fairness and sustainability, Islamic science can help to address economic inequalities and promote a more just and equitable society.

In conclusion, Islamic science has a long history of promoting social justice, with Islamic teachings and principles providing a framework for understanding and addressing social inequalities and injustices. By drawing on these principles, Islamic science can help to promote social justice in the modern world and contribute to the development of a more just and equitable society.

Chapter 48: Islamic Science and Education

Islamic science has had a profound impact on education, both historically and in the modern world. Islamic teachings emphasize the importance of seeking knowledge and understanding the natural world, and Islamic scholars and scientists have made significant contributions to the development of educational systems and practices.

Historically, Islamic scholars and scientists played a key role in the preservation and transmission of knowledge from ancient civilizations to the Islamic world and beyond. Islamic scholars translated and preserved Greek, Roman, Persian, Indian, and Chinese texts, making them accessible to scholars in Europe and other parts of the world. This transmission of knowledge helped to spark a revival of learning in Europe during the Renaissance and laid the foundation for modern education systems.

Islamic science also made important advancements in the field of education, developing innovative pedagogical techniques and methods. Islamic scholars and scientists developed the concept of the madrasa, or Islamic school, which provided education in a wide range of subjects, including theology, law, mathematics, astronomy, and medicine. The madrasa system emphasized the importance of holistic education, with a focus on both religious and secular subjects.

In the modern world, Islamic science continues to play a role in education, with Islamic teachings and principles informing educational practices in many Islamic countries. Islamic education systems often emphasize the importance of moral and ethical development, as well as the acquisition of knowledge and skills. Islamic schools and universities are also known for their strong emphasis on memorization and recitation of the Quran, which is considered a form of spiritual and intellectual development.

Islamic science has also influenced education in the broader sense, with Islamic scholars and scientists contributing to advancements in fields such as mathematics, astronomy, medicine, and philosophy. Many of the principles and practices developed by Islamic scholars and scientists continue to inform modern educational theories and practices, shaping the way we think about teaching and learning.

In conclusion, Islamic science has had a profound impact on education, both historically and in the modern world. By emphasizing the importance of seeking knowledge and understanding the natural world, Islamic science has helped to shape educational systems and practices that continue to influence education around the world.

Chapter 49: Islamic Science and the Quest for Knowledge

Islamic science has a deep-rooted tradition of seeking knowledge and understanding the natural world, guided by the teachings of Islam. The quest for knowledge is considered a fundamental duty in Islam, and Islamic scholars and scientists have made significant contributions to a wide range of fields, including astronomy, mathematics, medicine, and philosophy.

One of the key principles of Islamic science is the belief in the unity of knowledge. Islamic scholars and scientists believe that all knowledge is interconnected and that understanding one field of knowledge can lead to a greater understanding of the world as a whole. This holistic approach to knowledge has led to important advancements in fields such as astronomy, where Islamic astronomers developed models to explain the movements of celestial bodies, or in medicine, where Islamic physicians developed new treatments and surgical techniques based on their understanding of the human body.

Islamic science also emphasizes the importance of observation and experimentation. Islamic scholars and scientists have long recognized the importance of empirical evidence in understanding the natural world, and many of their advancements were based on careful observation and experimentation. For example, Islamic astronomers made accurate observations of the stars and planets, leading to new discoveries in astronomy, while Islamic physicians conducted experiments to test the effectiveness of different treatments and remedies.

Another key principle of Islamic science is the importance of ethical behavior and responsibility. Islamic scholars and scientists are guided by ethical principles that promote honesty, integrity, and compassion in their pursuit of knowledge. These ethical principles are reflected in the way Islamic scholars and scientists conduct their research and interact with others, leading to a more ethical and responsible approach to science.

In the modern world, Islamic science continues to inspire and guide the quest for knowledge. Islamic scholars and scientists are active in a wide range of fields, contributing to advancements in science, technology, and medicine. By drawing on the principles and values of Islamic science, researchers and scholars can continue to make valuable contributions to the quest for knowledge and understanding of the natural world.

Chapter 50: Conclusion: A Vision for the Future of Science in Islamic Civilization

The rich tradition of Islamic science has a promising future, with opportunities for further innovation, collaboration, and advancement. As we look ahead, we envision a future where Islamic science continues to thrive, making valuable contributions to the global scientific community and addressing some of the most pressing challenges facing humanity.

One of the key pillars of our vision for the future of Islamic science is continued investment in research and development. Islamic countries have made significant strides in recent years in investing in science and technology, and we believe that this trend will continue. By investing in research and development, Islamic countries can foster a culture of innovation and discovery that will drive advancements in science and technology.

Another key pillar of our vision is increased collaboration and exchange. Islamic science has a long history of collaboration with scholars and scientists from different cultures and backgrounds, and we believe that this tradition of collaboration should continue. By working together, researchers and scholars from Islamic and non-Islamic societies can share ideas, knowledge, and expertise, leading to new discoveries and innovations.

Education is also a key component of our vision for the future of Islamic science. Islamic teachings emphasize the importance of seeking knowledge and understanding the natural world, and we believe that education should be a priority in Islamic countries. By investing in education, Islamic countries can empower the next generation of scientists and scholars to make meaningful contributions to the field of Islamic science.

Finally, our vision for the future of Islamic science includes a commitment to ethical and responsible conduct. Islamic teachings emphasize the importance of ethical behavior and responsibility, and we believe that these principles should guide the practice of science. By conducting research and innovation ethically and responsibly, Islamic scientists and scholars can ensure that their work benefits humanity and contributes to the greater good.

In conclusion, the future of Islamic science is bright, with opportunities for further innovation, collaboration, and advancement. By investing in research and development, fostering collaboration and exchange, prioritizing education, and committing to ethical and responsible conduct, Islamic science can continue to thrive and make valuable contributions to the global scientific community.

About the author Binish Shah

Binish Shah Expertise in sales management extends globally, where she crafts strategic approaches for various international companies. Her role involves devising tailored strategies for companies traversing the globe, leveraging her extensive sales experience and understanding of diverse markets.

Beyond her professional prowess, Binish Shah passion for personal development shines. She embraces mindfulness, honing its practical applications for enhanced focus and mental well-being. Her journey includes conquering stage fright, mastering public speaking, and fostering personal growth.

Financially astute, Binish Shah adeptly manages personal finances, drawing on her understanding of influence and persuasion across both professional and personal spheres.

Despite a busy schedule, maintaining a healthy lifestyle remains paramount to Binish Shah. She not only creates nutritious meals swiftly but also explores meditation practices for inner tranquility.

Emphasizing healthy relationships through effective communication and boundaries, Binish Shah values continual self-reflection and personal growth.

Networking stands as a cornerstone for her career advancement. She refines time management skills to maximize productivity and attain her objectives.

Her grasp of investment fundamentals enables informed financial decisions. To maintain equilibrium, Binish Shah delves into stress reduction techniques like mindfulness, yoga, and meditation, fostering a balanced mindset.

Yours Sincerely

Binish Shah

Email: binish728@gmail.com

www.ingramcontent.com/pod-product-compliance
Lightning Source LLC
Chambersburg PA
CBHW050317230526
45471CB00005B/2233